MANAGING MAINTENANCE ERROR

We have left undone those thinges whiche we ought to have done, and we have done those thinges which we ought not to have done.

Book of Common Prayer, 1559

This book is dedicated to
John Goglia, *NTSB*
Chow Hock Lin, *Singapore Airlines Engineering Company*
Mike Innes, *Cathay Pacific*
Maintenance Crew *S11*

MANAGING MAINTENANCE ERROR

A Practical Guide

JAMES REASON
ALAN HOBBS

ASHGATE

Published by
Ashgate Publishing Company
Gower House
Croft Road
Aldershot
Hampshire GU11 3HR
England

Ashgate Publishing Company
Suite 420
101 Cherry Street
Burlington, VT 05401-4405
USA

Ashgate website: http://www.ashgate.com

British Library Cataloguing in Publication Data
Reason, James
Managing maintenance error : a practical guide
1.Airplanes - Maintenance and repair 2.Maintenance -
Management 3.Repairing - Management 4.Errors - Prevention
I.Title II.Hobbs, Alan
629.1'346'068

Library of Congress Cataloging-in-Publication Data
Reason, J.T.
Managing maintenance error : a practical guide/James Reason, Alan Hobbs.
p. cm.
Includes bibliographical references and index.
ISBN 0-7546-1590-1 (alk. paper)--ISBN 0-7546-1591-X (pbk: alk paper)
1. Plant maintenance. 2. Industrial equipment--Maintenance and repair. 3. Human engineering. I. Hobbs, Alan, 1962- II. Title

TS192 .R42 2002
658.2'02--dc21

2002024926

ISBN 0 7546 1590 1 (Hbk)
ISBN 0 7546 1591 X (Pbk)

Typeset by Manton Typesetters, Louth, Lincolnshire, UK.
Printed and bound in Great Britain by MPG Books Ltd, Bodmin, Cornwall.

Contents

List of Figures

Figure 6.1 with kind permission of Photovault.
Figures 6.2 and 6.3 with kind permission of British Transport Police.

List of Tables

Preface

Imagine a phenomenon that has created huge financial losses each year and, worse, resulted in death and injury throughout the world. Suppose that this poorly understood hazard caused aircraft to crash, medical equipment to malfunction, and crucial technical systems to fail at inopportune times. Under these circumstances, we would expect a worldwide effort to understand the problem and counteract it.

As it turns out there *is* such a phenomenon, yet it receives little attention and rarely makes the headlines. We are referring to maintenance error. While human fallibility is not new, it is becoming painfully apparent that modern technology can amplify its effects. Three hundred years ago a maintenance error, say in fitting a horseshoe, might have produced consequences that would affect a handful of people at most. Today, the implications of maintenance errors can reach further and cause far greater harm than ever before, leading to aircraft accidents and railway disasters that can claim hundreds of lives, oil spills causing long-term environmental damage, or potentially catastrophic accidents in the chemical and nuclear industries.

In many industries, safety and reliability have been improved by the automation of tasks previously performed by humans. But maintenance is not so easy to automate. So long as we rely on human hands and minds to maintain our technology, we face the irony that maintenance is a significant—some say the major—cause of failures. There are two maintenance-related threats to the integrity of systems. The first is that someone will not perform or will fail to complete a maintenance task designed to address an actual or potential failure, or will not detect signs of a failure. The second risk is that maintenance personnel will introduce the ingredients for a failure that would not have otherwise occurred.

The central contention of this book is that while the risk of maintenance error can never be eliminated entirely, it can be managed more effectively. Maintenance personnel and their managers need to understand why maintenance errors occur, and how the risk of error can be controlled. Most of our case studies are drawn from the aviation industry, but as our focus is on the human element of maintenance rather than the specific asset being maintained, we are confident that this book will be useful to all kinds of maintainers.

Our target readership is people who manage, supervise or carry out maintenance activities in a wide range of industries. Our main aim has been to provide a set of basic principles that can be adapted to your local needs. We discuss a variety of error management techniques, but this book is not intended to provide a comprehensive inventory of available, off-the-shelf tools. Rather, our goal has been to create a mindset from which effective remedies can be created to suit your particular conditions. It is our experience that maintenance personnel are ingenious and adaptable people who are perfectly capable of devising their own solutions. There is no one best way, and more than half the battle is grappling with the basic issues.

Although both authors are human factors specialists, we have tried to avoid psychobabble as far as possible. Both of us have spent a considerable time observing the work of maintenance facilities, and talking with maintainers—albeit mostly in the aviation context—and we believe we have a fair understanding of the nature of maintenance work and of the pressures under which it is carried out. But for those occasions where we have failed to bridge the disciplinary gap adequately, we offer our apologies in advance. Being human, we too are fallible.

Finally, we would like to acknowledge our considerable gratitude to all those maintainers—particularly in the United Kingdom, the United States, Singapore and Australia—who tolerated our presence so amiably and who gave of their valuable time and expertise to explain the complexities of their craft. We would also like to thank our human factors colleagues who provided valuable comments on various draft versions.

James Reason
Alan Hobbs

1 Human Performance Problems in Maintenance

The Bad News

If some evil genius were given the job of creating an activity guaranteed to produce an abundance of errors, he or she would probably come up with something that involved the frequent removal and replacement of large numbers of varied components, often carried out in cramped and poorly lit spaces with less-than-adequate tools, and usually under severe time pressure. There could also be some additional refinements. Thus, it could be arranged that the people who wrote the manuals and procedures rarely if ever carried out the activity under real-life conditions. It could also be decreed that those who started a job need not necessarily be the ones required to finish it. A further twist might be that a number of different groups work on the same item of equipment either simultaneously or sequentially, or both together.

Small wonder, then, that maintenance-related activities attract far more than their fair share of human performance problems. Table 1.1

Table 1.1 The relationships between activities and performance problems in nuclear power plant events*

Type of activity	Proportions of human performance problems associated with activity type
Maintenance, calibration and testing	Range = 42–65%
Normal plant operations	Range = 8–30%
Abnormal and emergency operations	Range = 1–8%

* Combined results from three US and one Japanese survey (see note 1)

shows the combined results of four surveys, three relating to US nuclear power plants (NPPs) and one dealing with Japanese NPPs.[1] It can be seen that the proportions of human performance problems associated with maintenance-related activities far exceeded those relating to other kinds of human performance. In three out of four of these studies, maintenance errors accounted for more than half of all the root causes of potentially serious events. Comparable data are not available for other safety-critical industries, but the fact that all maintenance-related tasks have a great deal in common suggests that such numbers would not be wildly dissimilar to those observed in nuclear power generation.

Maintenance errors have been among the principal causes of several major accidents in a wide range of technologies. These include:

- the Apollo 13 oxygen tank blow out (1970)
- the Flixborough cyclohexane explosion (1974)
- the loss of coolant near-disaster at the Three Mile Island nuclear power plant in Pennsylvania (1979)
- the crash of a DC10 at Chicago O'Hare (1979)
- the calamitous discharge of methyl isocyanate at a pesticide plant near the Indian city of Bhopal (1984)
- the crash of a Japan Air Lines B747 into the side of Mount Osutaka (1985)
- the explosion on the *Piper Alpha* oil and gas platform in the North Sea (1988)
- the Clapham Junction rail collision (1988)
- the explosion at the Phillips 66 Houston Chemical Complex in Pasadena, Texas (1989)
- the blow out of a flight deck window on a BAC1–11 over Oxfordshire (1990)
- the in-flight structural break of an Embraer 120 at Eagle Lake, Texas (1991)
- a blocked pitot tube contributing to the total loss of a B757 at Puerto Plata in the Dominican Republic (1996)
- the oxygen generator fire in the hold of a DC9 over Florida (1996).

It has also been estimated that maintenance errors ranked second only to controlled flight into terrain accidents in causing onboard aircraft fatalities between 1982 and 1991.[2]

Despite these tragedies, the main repercussions of maintenance errors are more likely to be felt in the bottom line than in injuries and fatalities. Maintenance errors cause large and continuing financial losses. Maintenance-induced or maintenance-prolonged outages in US nuclear power plants are priced at around a million dollars a day.

At coal-fired power stations, 56 per cent of forced outages occur less than a week after a planned or maintenance shutdown.[3] General Electric has estimated that each in-flight engine shutdown—for which maintenance errors are usually the primary cause—costs the airline in the region of $500 000. Boeing rates the costs of each maintenance-related flight cancellation at $50 000, and at $10–20 000 for each hour of maintenance-induced delay.[4] Litigation costs related to the ValuJet DC9 crash in the Florida Everglades are currently in excess of one billion dollars. And so it goes on. Maintenance errors not only endanger lives and assets, they are also extremely bad for business. Yet they keep on happening in remarkably similar ways—which brings us to the good news.

The Good News

Many people regard errors as random occurrences, events that are so wayward and unpredictable as to be beyond effective control. But this is not the case. While it is true that chance factors play their part and that human fallibility will never be wholly eliminated, the large majority of slips, lapses and mistakes fall into systematic and recurrent patterns. And these patterns are especially evident in maintenance-related activities, as we shall see below.

Far from being entirely unpredictable happenings, maintenance mishaps fall mostly into well-defined clusters shaped largely by situation and task factors that are common to maintenance activities in general. That these errors are not committed by a few careless or incompetent individuals is evident from the way that different people in different kinds of maintenance organisations—often very good people in excellent organisations—keep on making the same blunders. One of the basic principles of error management is that the best people can make the worst mistakes.

In 1997 Alan Hobbs interviewed experienced aircraft maintainers about incidents they had been involved in, or had witnessed, and was told of 86 safety incidents. Human error featured in all but a few of these events. About half of the incidents had implications for worker safety and about half affected the airworthiness of aircraft. When the people who supplied these incident reports were asked to say, in each case, whether or not a similar or identical incident had happened before, they produced the responses summarized in Figure 1.1.[5]

More than half of the incidents (particularly those with bad consequences for the aircraft) were recognized as having happened before. In the majority of cases the maintainers making these judgements were confident that the same or similar errors could happen again.

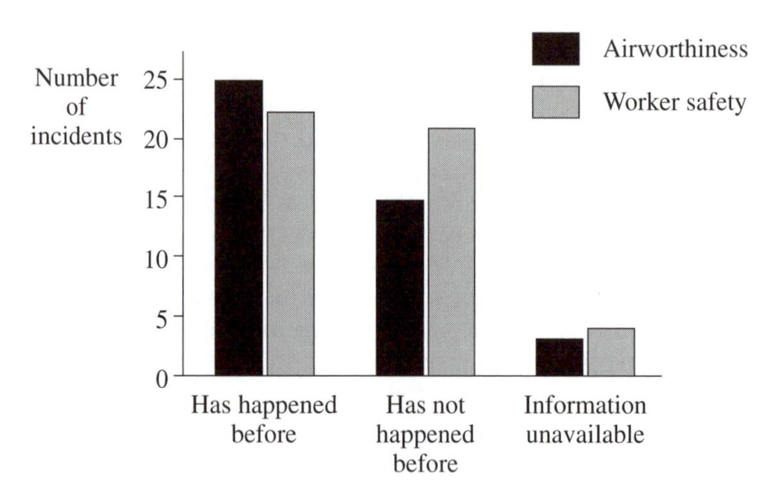

Figure 1.1 The frequency of reoccurence of aircraft maintenance incidents (n = 86)

Source: A. Hobbs *Human Factors in Airline Maintenance: A Study of Incident Reports* (Canberra: Bureau of Air Safety Investigation, 1997).

This is an important finding because it shows that there are certain situations and work pressures that lead people into the same kind of error regardless of who is doing the job. These 'error traps' clearly imply that we are dealing primarily with error-provoking tasks and error-inducing situations rather than with error-prone people.

So the good news boils down to this: the maintenance error problem can be managed in the same way that any well-defined business risk can be managed. And because most maintenance errors occur as recognizable and recurrent types, limited resources can be targeted to achieve maximum remedial effect. It should be stressed, however, that there is no one best way of limiting and containing human error. As discussed in Chapter 2, effective error management requires a wide variety of counter-measures directed at different levels of the system: the individual, the team, the task, the workplace and the organization as a whole. First, however, we will look at the patterns into which maintenance errors fall.

Removal Versus Replacement

Regardless of the industry or sphere of operations, many maintenance activities involve two repeated activities: (a) the removal of fastenings together with the disassembly of components, and (b) their reassembly and installation followed by the replacement of the fastenings. Anyone who has ever taken anything apart and then tried to put the bits back together again knows that the former is far easier to accomplish than the latter. There is often something left over when you think you have completed the reassembly.

It does not take a rocket scientist to work out why reassembly is more vulnerable to human error than disassembly. Consider the following example as a micro-model of maintenance in general. Imagine a bolt with eight nuts on it. The nuts are labelled A–H and they need to be removed and then replaced in a predetermined order. There is really only one way of taking the nuts off the bolt, and each step is prompted naturally by the preceding one. All the necessary knowledge is available in the task itself and does not have to be stored in memory or read from procedures. It is 'knowledge-in-the-world' rather than 'knowledge-in-the-head'.

However, when it comes to reassembling the nuts in a particular order, there are over 40 000 ways of getting the sequence wrong (factorial 8 possible combinations: $8 \times 7 \times 6 \times 5 \times 4 \times 3 \times 2 \times 1 = 40\,320$)—and this takes no account of any possible omissions. Moreover, all the necessary knowledge has to be either contained in memory or available on some written procedure. Reassembly imposes a much greater burden upon limited mental resources (that is, memory and attention), thus greatly increasing the probability of error.

In a real-life assembly task (as opposed to the bolt-and-nuts example), the maintainer's job is made even more difficult by the fact that errors (omissions, improper installations, misorderings and so on) may be covered up by the installation process. Thus, installation and reassembly suffer from a double disadvantage: not only is the probability of making an error much greater than during disassembly, the chances of detecting it are very much less.

There is now a wealth of evidence, mainly from aviation, to show that by far the largest number of maintenance errors is associated with reassembly and installation. Analysis carried out by Boeing, for example, identified the top seven causes of in-flight engine shutdowns (IFSDs) as follows—the numbers in parentheses show the percentage of each causal category as a proportion of all causal categories:[6]

- Incomplete installation (33%)
- Damaged on installation (14.5%)
- Improper installation (11%)
- Equipment not installed or missing (11%)
- Foreign object damage (6.5%)
- Improper fault isolation, inspection, test (6%)
- Equipment not activated or deactivated (4%).

Pratt and Whitney in a 1992 survey of 120 IFSDs occurring on B747s in 1991 obtained virtually identical findings.[7] The top three contributing factors were missing parts, incorrect parts and incorrect installation. The UK Civil Aviation Authority surveyed maintenance deficiencies of all kinds and found that the commonest problem was the incorrect installation of components, followed by the fitting of wrong parts, electrical wiring discrepancies and foreign objects (tools, rags, personal items) left in the aircraft after maintenance.[8]

Commission Versus Omission Errors

Put very simply, there are two ways people can go wrong. They can either do something they should not have done, or fail to do something they should have done. The former are errors of commission, the latter are errors of omission. Here we will consider the evidence showing that omissions—failures to carry out necessary actions or tasks, usually during installation—are the largest single category of maintenance error. Indeed, in nuclear power generation, and possibly elsewhere as well, omission errors during maintenance make up the single largest class of human performance problems in the system as a whole—where this includes errors made during normal control operations as well as while recovering from an emergency or off-normal state.

An analysis of 200 significant event reports from nuclear power plants identified the omission of functionally isolated acts as accounting for 34 per cent of errors recorded—the largest single category. The same study also showed that the activities most often associated with these omission errors were repair and modification (41 per cent), testing and calibration (33 per cent), inventory control (9 per cent) and manual operation and control (6 per cent).[9]

Another investigation involving US nuclear power plant operations found that 64.5 per cent of the errors associated with maintenance-related activities involved the omission of necessary steps.[10] A very similar proportion for maintenance omissions was found in a survey of Japanese nuclear power plants.[11]

Comparable figures are found in aviation maintenance. In one study, based on an analysis of 122 maintenance errors recorded by a major UK airline over a three-year period, omissions accounted for 56 per cent of the total, 30 per cent involved incorrect installations of one kind or another, while 8 per cent involved using the wrong parts.[12] When these omission errors were examined in detail, the following sub-categories were found:

- Fastenings left undone or incomplete (22%)
- Items left locked or pins not removed (13%)
- Caps loose or missing (11%)
- Items left loose or disconnected (10%)
- Items missing (10%)
- Tools and/or spare fastenings not removed (10%)
- Lack of lubrication (7%)
- Panels left off (3%)
- Other (14%).

In the series of interviews with experienced maintainers referred to earlier, Alan Hobbs was told of 120 errors that led to incidents.[13] The largest single category was omissions, which accounted for 48 per cent of all errors. The various sub-categories of omission errors are listed below in order of frequency of occurrence:

- Aircraft system not locked out or made safe before starting work
- Incomplete installation
- Work not documented
- System not reactivated or deactivated
- Verbal warning not given
- Pin or tie left in place
- Warning sign or tag not used
- Incomplete or inadequate testing
- Material left in aircraft or engine
- Access panel not closed
- Equipment not installed
- Required servicing not done.

The point is made. Omissions of one kind or another make up the largest single category of maintenance errors. Such errors are commonly associated with reassembly and installation activities. Naturally enough, these errors involve those items or actions that lend themselves most readily to being omitted: fastenings, pins and caps left off or left loose; foreign objects not removed; warnings not given; paperwork neglected. In Chapter 9, we will consider the factors that are most influential in provoking omission errors.

Summary

The chapter began with the 'bad news' that maintenance is a highly error-productive activity. It attracts a large—perhaps the largest—proportion of human factors problems in a wide range of hazardous technologies. The 'good news' is that maintenance errors are not random. They fall into systematic patterns relating to both the nature of the activity and the types of error involved. There is considerable evidence to show that reassembly and installation are associated with the lion's share of the errors made. In addition, omissions—failing to carry out necessary actions, usually when putting things back—make up the largest single category of maintenance errors. Finally, most of these errors have been judged by experienced maintainers as having happened before—and were also seen as likely to happen again. The fact that the same errors keep on happening to different people in different organizations strongly suggests that we should focus our remedial attention more upon the task and the workplace than upon the presumed psychological inadequacies of those making the errors. This has very important implications for managing maintenance error, as we shall see in the following chapters.

Notes

1 INPO, *An Analysis of Root Causes in 1983 Significant Event Reports*, INPO 84-027 (Atlanta, GA: Institute of Nuclear Power Operations, 1984); and INPO, *An Analysis of Root Causes in 1983 and 1984 Significant Event Reports*, INPO 85-027 (Atlanta, GA: Institute of Nuclear Power Operations, 1985); K. Takano, Personal communication, 1996.
2 R.A. Davis, 'Human factors in the global market place', Keynote Address, Annual Meeting of the Human Factors and Ergonomics Society, Seattle, WA, 12 October 1993. See also Boeing, *Statistical Summary of Commercial Jet Aircraft Accidents, 1959–92* (Seattle, WA: Boeing Commercial Airplane Group, 1993).
3 A. Smith, *Reliability-Centered Maintenance* (Boston: McGraw Hill, 1992).
4 R.C. Graeber, 'The value of human factors awareness for airline management', Paper presented to conference on Human Factors for Aerospace Leaders, Royal Aeronautical Society, London, 28 May 1996; D.A. Marx and R.C. Graeber, 'Human error in aircraft maintenance', in N. Johnston, N. MacDonald and R. Fuller (eds), *Aviation Psychology in Practice* (Aldershot: Avebury, 1994).
5 A. Hobbs, *Human Factors in Aircraft Maintenance: A Study of Incident Reports* (Canberra: Bureau of Air Safety Investigation, 1997).
6 Boeing, *Maintenance Error Decision Aid* (Seattle, WA: Boeing Commercial Airplane Group, 1994).
7 Pratt and Whitney, *Open Cowl*, March issue, 1992.
8 United Kingdom Civil Aviation Authority (UK CAA), 'Maintenance error', *Asia Pacific Air Safety*, September 1992.
9 J. Rasmussen, 'What can be learned from human error reports?', in K. Duncan, M. Gruneberg and D. Wallis (eds), *Changes in Working Life* (London: Wiley, 1980).
10 INPO 1984 and INPO 1985, op. cit.
11 Takano, op. cit.
12 J. Reason, *Comprehensive Error Management in Aircraft Engineering: A Manager's Guide* (Heathrow: British Airways Engineering, 1995).
13 Hobbs, op. cit.

2 The Human Risks

Taking a Systems View

> A routine six-monthly performance inspection of an Uninterruptible Power Supply (UPS) for the air traffic control centre at a major city airport was commenced at 6.00 pm on a busy weekday. Approximately 20 minutes after work commenced, the centre sustained a total loss of electrical power. As a result, air traffic control screens failed, software switching of voice communications channels was interrupted, as were satellite communications, the radar feeds to two other major cities were interrupted and the lights in the control room went out. Air traffic controllers were unable to determine the positions of aircraft for about 7 to 10 minutes. By using the emergency radio, controllers were able to direct flight crews to keep a visual lookout for aircraft and to ensure that their Traffic Alert and Collision Avoidance Systems were switched on.[1]

When confronting the realities of maintenance error on a daily basis, it is very easy to get tangled up in the local details of each incident and to lose sight of the broader picture. We get irritated with the error maker and wonder how anyone could have been so irresponsible, so careless or so stupid, particularly when any kind of lapse on the part of a maintainer could have very damaging consequences. These are natural and understandable reactions, but they can lead maintenance managers to some fundamentally wrong conclusions about how to deal with the error problem.

The management of error is itself prone to error. It seems obvious that a human error must have human origins. Some individual—or, quite often, a group of individuals—went wrong and, in so doing, endangered the safety of the operation and the lives of third parties. The temptation, then, is to home in upon the individual psychological factors immediately preceding the making of an error and to do whatever seems necessary to prevent their recurrence. Such measures usually take the form of disciplinary action, writing another procedure, blaming, shaming and retraining. But this is to miss two important points about human error. First, errors are inevitable. Everyone makes them, but no one actually intends them to happen. Second, errors are consequences not just causes. They do not occur as isolated glitches in people's minds. Rather, they are shaped by local circumstances: by the task, the tools and equipment and the workplace in general. If we are to understand the significance of these contextual factors, we have to stand back from what went on in the error maker's head and consider the nature of the system as a whole. If this book has a constant theme it is that situations and systems are easier to change than the human condition—particularly when the people concerned are already well-trained and well-motivated, as they usually are in maintenance organizations.

Systems with Human Elements

Maintenance organizations are systems. A system is a set of interrelated entities or elements. Systems can be either natural (for example, the solar system) or man-made. In the latter, the constituent elements can be—and usually are—both people and technical objects. Man-made systems are organized for a particular purpose and tend to be bounded—though the boundary is usually permeable. A system with human elements is shown symbolically in Figure 2.1.

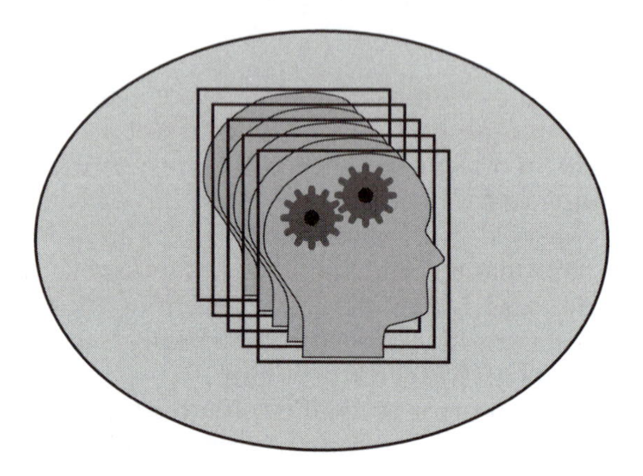

Figure 2.1 A system with human elements

Systems with human elements take many different forms. Listed below are examples of man-made systems with human elements:

- A large city
- A road network
- A surgical centre
- A football stadium
- A bank dealing room
- A public administration
- A maintenance organization.

Human-related Disturbances

All systems with people in them suffer human-related disturbances. These are unwanted deviations from some desired norm that have their roots in various kinds of human behavioural tendencies. We can represent these human-related disturbances by the 'out-of-limits' symbol shown in Figure 2.2.

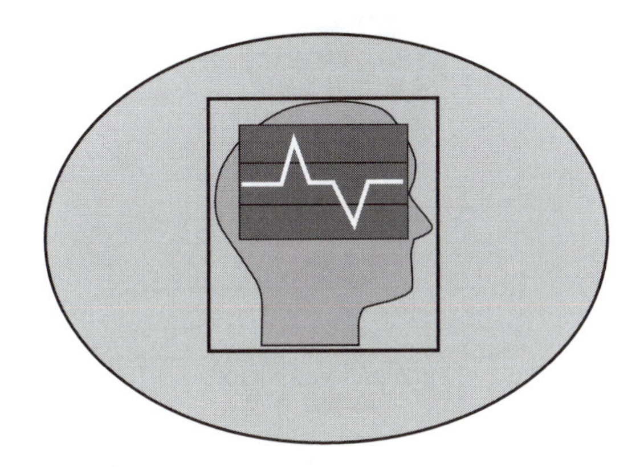

Figure 2.2 A symbolic representation of a human-related disturbance

Human-related disturbances can take many different forms, but each system tends to have some predominant type(s). For example:

- In large cities, it is crimes against the person and theft, often associated with drug use or trading.
- On the roads, it takes the form of dangerous or reckless driving (most particularly by young men).
- In surgical centres, it could be lack of surgical skill in regard to a particular procedure.
- In football stadiums, it is the physical press of large crowds

combined with the risk of violent clashes between supporters (for example, Heysel and Hillsborough).

- In bank dealing rooms, it could be rogue traders (for example, the Barings Bank collapse).
- In public administrations, it may be corrupt officials (instances too numerous to name).
- In maintenance organizations, it is human error.

Each Disturbance has a History

As argued above, human-related disturbances do not just emerge ready-made from people's heads. They have a history within the system. That is, they are the consequence of many interacting contextual factors. This is shown diagrammatically in Figure 2.3.

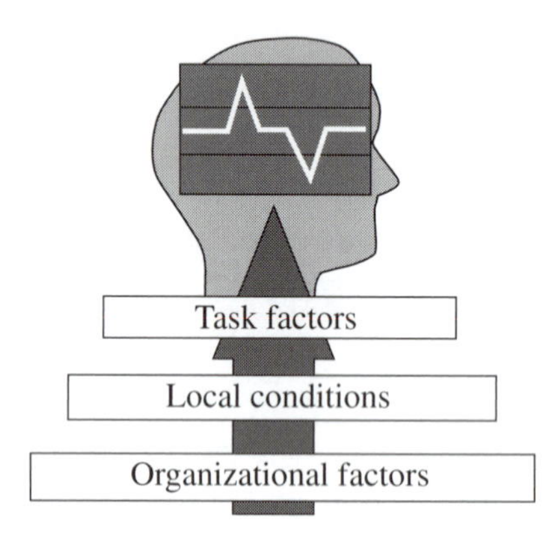

Figure 2.3 Factors involved in the occurrence of human-related disturbances

Both the nature and the likelihood of disturbances depend upon what was being done at the time, the surrounding circumstances and the character of the system as a whole. Exactly how these factors can affect human-related disturbances, such as errors, will be considered in detail in later chapters.

Systems Build Defences Against Foreseeable Disturbances

In most established systems, the nature of the predominant disturbances is well known and, to a large extent, foreseeable. As a consequence, defences can be created to protect against these expected disturbances. The provision of these barriers and safeguards is the responsibility of those who manage or control the system. Experience has shown that no one type of defence is sufficient—for the reasons discussed below. It is therefore more usual to rely upon defences-in-depth. That is, many layers of protection located between the disturbance and the vulnerable parts of the system. Ideally, defences should possess both redundancy (multiple backups) and diversity (a variety of different safeguards). Such a defensive arrangement is typical of most modern technological systems and is illustrated in Figure 2.4.[2]

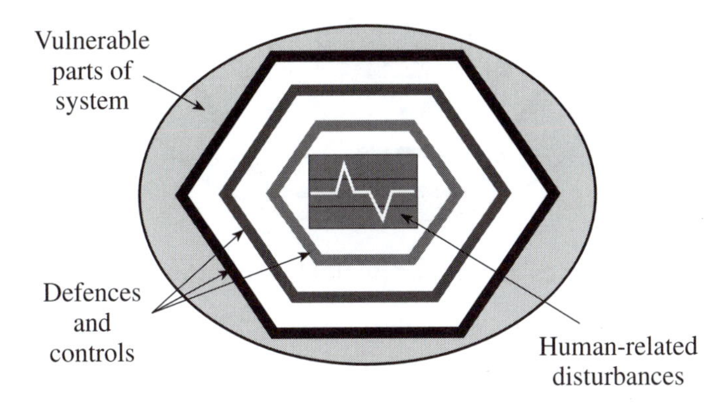

Figure 2.4 A many-layered defensive arrangement intervenes between the disturbance and the vulnerable parts of the system

The composition of any set of defences depends upon the nature of the system. In large cities, for example, defences will comprise law enforcement agencies, the courts and penal institutions. In maintenance organizations, however, they are made up of a mixture of 'hard' and 'soft' defences. Hard defences involve a variety of engineered safety features: physical safety barriers, sensing devices, warnings and alarms. Soft defences—'soft' because they involve a mixture of people and paper-based measures—will include such controls as rules, regulations, procedures, supervisory checking and sign-offs, audits, shift handover procedures, permit-to-work systems and quality assurance personnel.

System Defences can also Fail

No one layer of defence will ever be entirely intact, as shown in Figure 2.5. There will always be holes and gaps. Defences are designed and implemented by fallible people. No one can foresee all possible combinations of disturbances.

Figure 2.5 Defences are never perfect: each one has holes and gaps

For some bad outcome or disaster to occur within a man-made system, two conditions must be fulfilled. First, there needs to be the occurrence of some initiating disturbance. Second, the system defences fail to detect the occurrence of the disturbance and/or fail to contain the bad consequences. For a disaster to happen, it is often necessary for many defensive layers to be breached at the same time. Since there are many defences, such bad outcomes are usually much fewer than the number of potentially dangerous disturbances.

It is often said in retrospect that a particular bad event was avoidable. However, this is true in only one sense. While the defensive failures may have been preventable, the human-related disturbances they were meant to protect against are not. Violence, recklessness, criminal behaviour and fallibility are deeply rooted in human psychology and are largely fixtures. They can be moderated, but never entirely eliminated.

The Moral Issue

To those with a commitment to the perfectibility of humankind, these views must seem unacceptably fatalistic. But this objection overlooks the difference between the individual person and the human race in general. Individuals may be changeable, but the human species is probably not. Evolution has dealt us a very mixed hand: combining a truly remarkable set of creative, communicative and constructional abilities with some fairly primitive animal drives and instincts.

It is in regard to human error, however, that these moral considerations are most likely to cloud and confuse the issue. There are two related problems. The first comes from failing to distinguish errors from other types of human-related disturbance. The second arises from assuming that errors are intrinsically bad things. When searching for the causes of some incident or accident, many system managers treat all human-related disturbances as reprehensible acts and punish them accordingly. Some of these human-related disturbances are clearly morally unacceptable—criminal behaviour, crowd violence and dangerous driving, for instance. But errors, in themselves, are not. They are quite another class of disturbance, being both universal and unintended. Nor are they intrinsically bad in the sense of being design defects in the human mind. They are merely the downside of having a brain. Each error tendency—inattention, forgetfulness, strong habit intrusions and the like—lies on the debit side of a mental balance sheet that stands very much in credit. Each recurrent error form is rooted in highly adaptive mental processes. Moreover, trial-and-error learning is an essential step in acquiring new skills or dealing with novel situations. It is not the error tendency that is the problem, but the relatively unforgiving and artificial (at least from an evolutionary perspective) circumstances in which errors are made. Switching on the toaster instead of the kettle in a kitchen may cause the perpetrator mild embarrassment, but exactly the same type of error committed in an aircraft maintenance hangar can have devastating consequences. The error is identical, only the situations differ.

Why are we so inclined to blame people rather than situations? Part of the answer has to do with what psychologists call the *fundamental attribution error*. This tendency is deeply rooted in human nature and relates to the differences in viewpoint between observers and actors. When we see or hear of someone making an error, we attribute this to the person's character or ability—to his or her personal qualities. We say that he or she was careless, silly, incompetent, reckless or thoughtless. But if you were to ask the person in question why the error happened, they would almost certainly tell you how the local circumstances forced them to act in that way. The truth, of course, lies somewhere in between.[3]

To break free of this 'blame cycle' we need to recognize that human actions are almost always constrained by factors beyond the person's immediate control, and to recognize that people find it difficult to avoid those actions that they never intended to commit in the first place. Of course, people can and do behave carelessly and stupidly. We all do at some time or another. But a stupid or careless act does not necessarily make a stupid or careless person. Everyone is capable of a wide range of behaviours, sometimes inspired and sometimes foolish, but mostly they lie somewhere in between. One of the basic principles of error management is that the best people can sometimes make the worst mistakes.

Because they are insufficiently aware of the inevitability of error and, at the same time, are culturally committed to blame and punishment, many maintenance organizations focus most of their limited resources on trying to prevent errors at the level of the individual maintainer. They fail to recognize the following:

- Measures relying heavily upon exhortations and sanctions have only very limited effectiveness. In many cases, they do more harm than good.
- Safety-significant errors happen at all levels of the system, not just at the 'sharp end'.
- Slips, lapses and mistakes are a product of a combination of causes in which the individual psychological factors—momentary inattention, forgetting, misjudgement and the like—are often the last and least manageable links in the 'error chain'.
- Incidents and accidents, especially in maintenance, are more often the result of error-prone situations than error-prone people.

As a result, they tend to neglect measures aimed at:

- reducing the error-provoking nature of the task, the team, the workplace and the organization.
- strengthening and improving defences to limit and contain the bad effects of those errors that will still occur.

In summary, many maintenance organizations try to change the human condition when they should be changing the conditions under which people work and should be treating errors as an expected and foreseeable part of maintenance work.

Errors are like Mosquitoes

You can try to deal with mosquitoes one by one. You can swat them and you can spray them, but they still keep coming. The only effective measures are, first, to drain the swamps in which they breed and, second, to use various proven defences—mosquito netting, mosquito repellents and quinine-based pills—to guard against malaria (the bad outcome of a mosquito bite).

In the case of maintenance errors, the 'swamps' are the task, team, workplace and organizational factors that provoke errors. The defences are the system safeguards and barriers that detect and recover errors before they can have a damaging result. These two processes—removing error-promoting situations and improving defences—form the two most important parts of an effective error management programme. Focusing the bulk of the organization's limited resources on isolated errors committed by individual maintainers is like swatting or spraying mosquitoes one by one.

Looking Ahead

The next chapter covers some of the fundamental design specifications of human beings and explains why maintenance-related activities can be particularly error-provoking. In order to manage the circumstances that create error, we need to appreciate both the varieties of human error and the factors that promote them. These issues are discussed in Chapters 4 and 5. Chapter 6 presents case studies of three systems failures. Chapter 7 spells out the basic principles of effective error management. Chapters 8–10 describe error management techniques directed at the individual, the team, the task, the workplace and the organization. Chapter 11 discusses the vital issue of safety culture, and Chapter 12 focuses on making it all happen—the management of error management.

Notes

1 Australian Transport Safety Bureau, Air Safety Occurrence Report 200002836 (Canberra: Australian Transport Safety Bureau, 2001).
2 See also the 'Swiss cheese' model as described in J. Reason, *Managing the Risks of Organizational Accidents* (Aldershot: Ashgate, 1997).
3 S.T. Fiske and S.E. Taylor, *Social Cognition* (Reading, MA: Addison-Wesley, 1984).

3 The Fundamentals of Human Performance

Psychology Meets Engineering

For many with an engineering background, psychology seems a soft, fuzzy, 'people' business quite remote from their own tough-minded and rigorous technical concerns. But appearances can be deceptive. There are many psychologists, particularly those interested in human cognition—the study of attention, memory, thought, problem solving and the control of action—that think like engineers and are far removed from the popular stereotype of the 'psychologist-as-shrink'. Cognitive psychologists tend to treat the mind as something akin to an information-processing machine. But unlike most machines, opening it up and examining its structure would not reveal the workings of the human mind. All that could be seen by the unaided eye would be a large greyish-pink, spaghetti-like lump of nerve tissue and blood vessels. Of course, modern brain scanning devices could reveal much more, but even then the link between structure and function remains fairly mysterious.

To be more precise, the claim that some psychologists think like engineers actually means that they think like control engineers and systems analysts rather than engineers of the more 'hands on' varieties. But for all that, there is considerably more common ground between engineering and psychology than is usually imagined.

This chapter describes the various levels of human performance in engineering terms—indeed, a control engineer first classified these levels. The three performance levels form a basis for categorizing the varieties of human error. Human error is often treated as a uniform collection of unwanted acts. In reality, errors fall into quite distinct types that require different kinds of remedial measures and occur at different levels of the maintenance organization. Understanding these differences is an essential precondition for effective error management. They are discussed in detail in Chapter 4. But first we need a very rough idea of how the mind controls our actions.

A 'Blueprint' of Mental Functioning

Figure 3.1 outlines the basic structural components of mental functioning. It is not a wiring diagram or a picture of brain anatomy, rather it is a very simplified 'systems representation' of the important elements and their inter-connections.

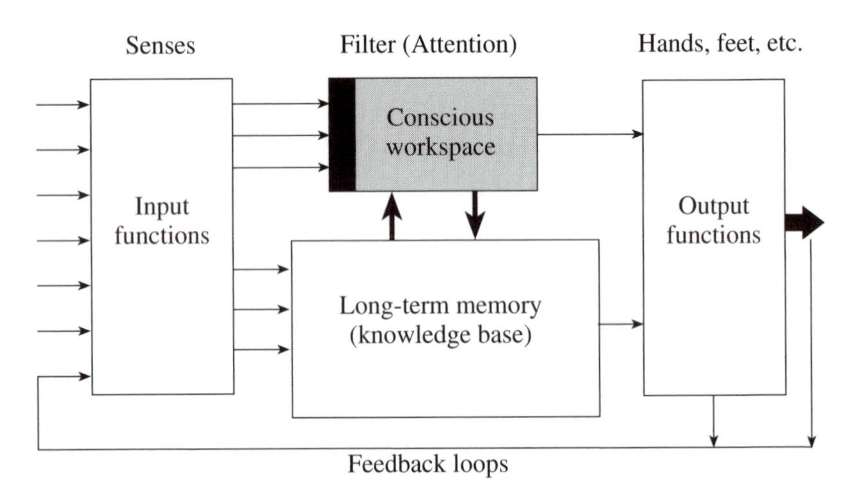

Figure 3.1 A simplified 'blueprint' of mental functioning

Sensory data comes into the brain via a number of input functions. These are the senses—vision, hearing, taste, touch, smell and the various position and motion receptors. Vision is the dominant sense, so much so that it can override conflicting inputs from the other senses.

A small proportion of the total available sense data enters the conscious workspace after having been screened by the attentional filter. Here, the selected data are broken down, added to or recombined by the computationally powerful (though slow and effortful) processes of thought, reasoning and judgement. The conscious workspace can act directly upon the various output functions—hands, limbs, feet, speech and so on—commanding them to perform certain actions or utterances.

Some of the input data passes directly to long-term memory, the knowledge base, where highly specialized knowledge structures or schemata act upon it. Whereas the conscious workspace keeps 'open house' to all kinds of sensory data, long-term memory seeks out only those pieces of information that are relevant to its stored knowledge structures. Each schema 'looks out for' only those bits of information relating specifically to it. Thus, a schema for recognizing cats is only interested in things having a certain size and shape, four legs, sharp claws, fur, whiskers and the like.

Information from long-term memory goes two ways.

- First, it can go directly to the output functions in the form of pre-packaged bursts of instructions. In engineering terms, we can say that whereas action control via the conscious workspace is *feedback-driven*, that via long-term memory is *feedforward-driven*. We will look at these control modes in more detail later.
- Second, the long-term memory is continually sending information to the conscious workspace. Sometimes, this information is deliberately sought by conscious retrieval mechanisms, but often items of information just pop into consciousness, seemingly of their own accord, but actually driven by the two basic search rules of *similarity matching* (match like to like) and *frequency gambling* (where two or more items have been located on the basis of similarity, favour the one that is most frequently and recently used in this situation).[1] An item of information thus delivered might be a thought, an image or an action.

The conscious workspace (CW) and long-term memory (LTM) usually operate in subtle harmony to guide our actions in the way that we intend. Sometimes things go wrong, but this happens comparatively rarely. The characteristics of these two controlling mechanisms (CW and LTM) stand in marked contrast one to another, as summarized in Table 3.1.

Limitations of the Conscious Workspace

As Table 3.1 indicates, an important difference between the conscious workspace and long-term memory is that the capacity of the conscious workspace is severely limited. When you look up a phone number and keep it in mind until you dial it, you are depending on the conscious workspace. Mental arithmetic is another situation where we have to keep several items stored in memory until we have found the answer. The conscious workspace has a time span of around one-and-a-half seconds, and operates like a leaky bucket. New bits of information or thoughts displace older items of information. The limits of the conscious workspace have important implications for maintenance work. Interruptions and other distractions can easily lead to steps being omitted and other failures of memory.

Table 3.1 Contrasting the properties of the conscious workspace and long-term memory

Conscious workspace properties	Long-term memory properties
A general problem solver	Vast collection of specialized 'experts'
Limited capacity	No limits yet established, either in terms of store size or duration of memories
Contents available to consciousness (i.e., we can describe them when asked)	Processes (though not the products) largely unconscious
Processes information sequentially (i.e., one thing at a time)	Processes information in parallel (i.e., many things going on at once)
Slow and laborious	Rapid and effortless
Essential for new tasks	Handles familiar routines and habits

Attention

What, in everyday language, we call 'attention' is very closely identified with the workings of the conscious workspace. Attention has a number of characteristics, as listed below.

- Attention is a limited commodity—if it is strongly drawn to one particular thing it is necessarily withdrawn from other competing concerns.
- These capacity limits give attention its selective properties—we can only attend to a very small proportion of the available sense data.
- Unrelated matters can capture attention—that is, preoccupation with some demanding sensory input or distraction by some current thoughts or worries.
- Attentional focus (concentration) is hard to sustain for more than a few seconds.
- The ability to concentrate depends very much upon the intrinsic interest of the current object of attention.
- The more skilled or the more habitual our actions, the less

attention they demand—indeed, skilled performance can be disrupted by having too much attention directed at it.
- Correct performance requires just the right balance of attention—neither too much nor too little.

We can summarize these features of attention by using an analogy: the torch beam model of attention. This is illustrated in Figure 3.2. Attention can be represented as a beam of light projected by a torch. We can portray the potential objects of attention as a task space. The torch illuminates only a very small area of this space. The present position of the spot of light represents that part of the task that is the current focus of attention. The torch can direct its beam to different areas of the task space.

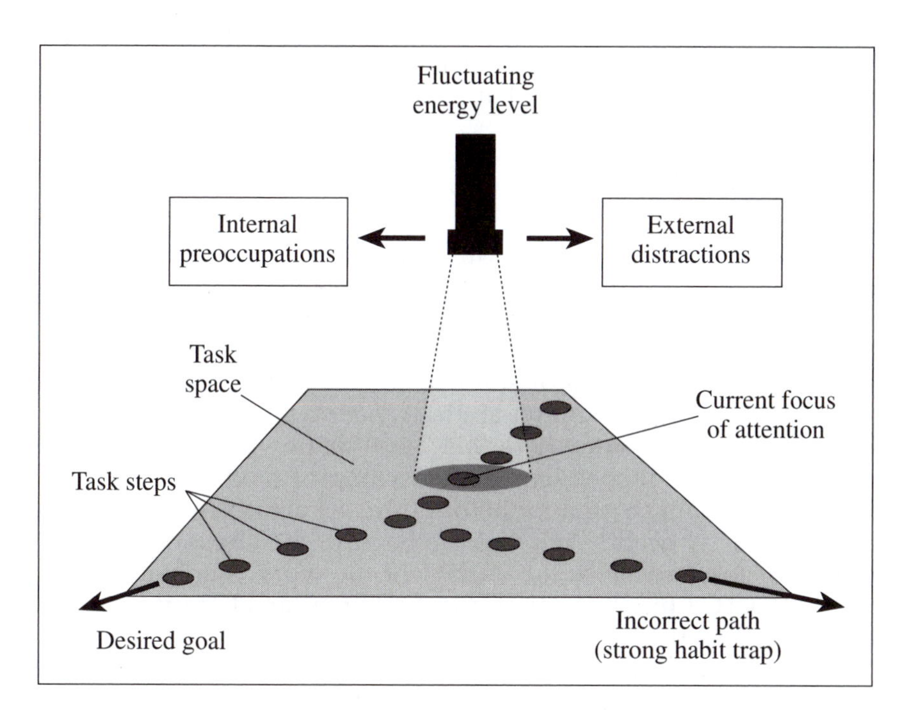

Figure 3.2 The torch beam model of human attention

A number of competing factors control the movements of the torch beam. First, and most obviously, there is the intention of the individual. We can direct the focus of our attention at will, but—as indicated above—we cannot necessarily sustain this attentional focus for very long, a few seconds at the most, before it is captured and redirected by some other concern. This is highly adaptive. Without this constant switching of attentional direction, we would have a keyhole view of the world. Many different things can claim attention. As shown in Figure 3.2, it can be grabbed by internal preoccupations or by external distractions. Any knowledge structure (or schema) in long-term memory can be fired up by emotion, or by its recency and frequency of use. The more fired up or activated the schema, the more likely it is to claim attention. This is fine when the schema in question is connected to the job in hand, but it could also relate to something quite different—a domestic worry or local distraction. This can result in a loss of concentration at a critical point in the task and produce an error. Figure 3.2 shows how a typical 'absent-minded' slip can occur. The initial steps in the task are common to two different onward paths, one of which has been travelled more frequently than the presently intended or correct route. Unless attention is directed to the choice point at the right moment, actions tend to follow the path of the strongest habit. These 'strong habit intrusions' are discussed further in the next chapter.

The Vigilance Decrement

In the Second World War, it was found that after about twenty minutes at their posts, radar operators became increasingly more likely to miss obvious targets. Often somebody walking by would casually notice a radar paint that had not been spotted by the operator, even though he or she was intently concentrating on the screen. This problem, known as the *vigilance decrement*, applies to many monitoring tasks where 'hits' are relatively few and far between. Boroscope inspections, the checking of medical X-rays and quality control inspection in factories are all areas where vigilance decrements may occur. Vigilance can be improved by increasing the number of rest breaks or the variety of the work. Vigilance is also often better in a more social atmosphere, perhaps because it keeps people more alert.

We will return to the torch beam model of attention in Chapter 8 when we consider the methods people can use to gain greater control of their attentional focus. In the meantime, we will go on to look at the relationship between attention and habit.

Attention and Habit

The more skilled or habitual our actions become, the less we need to guide the details of these actions by conscious attention. Habits (of thought and perception) and skilled actions are stored as pre-programmed strings of instructions in long-term memory. They are the 'software' of mental life.

The process of *automatization* (that is, actions being increasingly controlled by the 'automatic pilot') applies to all aspects of our lives—mental arithmetic, riding a bicycle, speaking a foreign language, dismantling a familiar piece of equipment, much of our thinking and even our social behaviour. The 'programs' stored in LTM guide the routine and recurrent actions of everyday living. This leaves the conscious workspace free to concentrate on broad strategy rather than tactical details, and to cope with new problems (that is, changes in the expected run of things) as they arise. Such devolution of control is essential for life; indeed, it is one of the human mind's greatest strengths. But, like most design features, it has a downside: the occasional appearance of actions-not-as-planned, or the absent-minded slips and lapses that will be discussed in Chapter 4.

One of the interesting features of acquiring a skill is that performance shifts from being directed by a largely verbal or 'digital' mode of control—talking yourself through the actions (or having an instructor do it)—to being guided by a mainly unconscious 'analogue' mode. There comes a point when a tennis coach or a flying instructor stops talking and says 'watch what I do' or 'follow me through on the controls'. Quite simply, skilled people are no longer able to describe their actions in words. This is called *the paradox of expertise*. If you were to try to tell an apprentice how to tighten a screw, you would soon find yourself making twisting movements with your right hand and saying 'do it like this'. The other side of the same coin is that if you concentrated too hard on the details of your habitual actions, you would almost certainly end up disrupting them. Try typing and thinking what the middle finger of your left hand is doing or, less wisely, running down stairs and asking yourself what your feet are doing.

Control Modes and Situations

How we behave is a complex function of two things: what is going on between our ears and what is happening around us. Much of cognitive psychology is concerned with trying to understand how these two primary influences interact to shape actions. As discussed above, human beings guide their actions by a variety of control modes. These are best represented as a continuum with two extremes and a mixture of the two in between, as shown in Figure 3.3.

At one extreme, there is the conscious mode of control. Here, we have to attend very closely to our actions and make deliberate and thought-out decisions about what we are going to do and how we

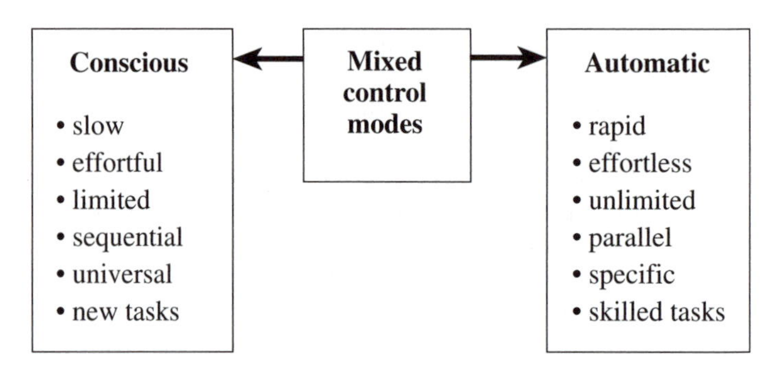

Figure 3.3 A continuum of action control

are going to do it. This control mode is slow, sequential, limited and highly error-prone (trial-and-error learning is an important means by which we acquire new skills), but computationally very powerful. In short, we can do all sorts of clever things with it if the conditions are right. It is essential for coping with new problems or new situations. But it is hard work and we do not really like it.

At the other extreme is the automatic control mode ('automatic pilot'). This is very fast and effortless—so we like it. We use it to control the routine, practised and highly skilled actions that we perform every day. When we work this way, we can (and often must) think about other things. Indeed, as mentioned above, too much conscious attention given to habitual acts can be very disruptive. Too little attention and our actions stray from our intentions along well-trodden but not currently intended pathways (more of this in Chapter 4).

In between, we switch between these two extreme modes, running off sequences of actions automatically, and then filling in the gaps by consciously thinking about what we are going to do next and checking on our progress. Most of the time, we work in this mixed-control way. It is what we do best. Only very rarely do we have to function entirely in the laborious conscious mode or in the unthinking automatic mode.

The situations in which we live and work can also be described by assigning them to various positions along a continuum. This is shown in Figure 3.4. At one extreme, there are highly familiar, routine and largely non-problematic situations. These are places like offices, workplaces, bathrooms, bedrooms, kitchens and all the other settings in which we carry out our recurrent everyday activities. In the middle, there are situations that pose expected or anticipated problems for which we have been trained, and for which we possess a whole stock of stored solutions in our long-term memories. At the

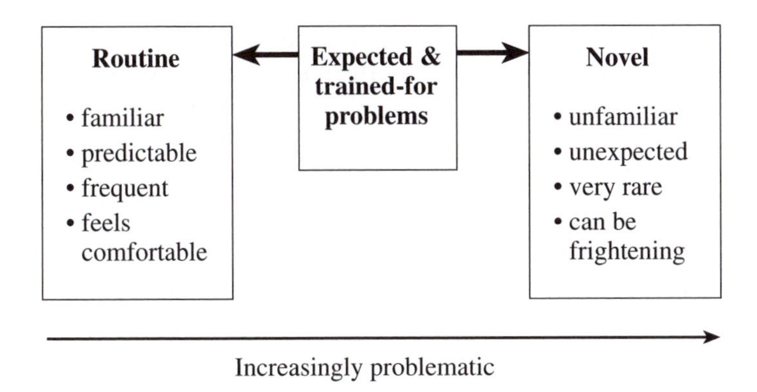

Figure 3.4 Classifying situations along a continuum

other extreme, there are entirely new situations that present difficult and often potentially dangerous problems. Here, we run out of pre-packaged solutions and have to find a way of resolving the difficulties on the spot.

Three Performance Levels

The distinguished Danish control engineer Jens Rasmussen did the worlds of both hazardous work and human factors a very great favour when he defined three levels of human performance in a way that was easily recognizable by engineers and readily acceptable to most cognitive psychologists (they never all agree about anything).[2] These levels are termed *skill-based* (SB), *rule-based* (RB) and *knowledge-based* (KB), and their main properties are summarized in Table 3.2. They have become something of an industry-standard in aviation, nuclear power generation, oil exploration and production, and in many other spheres that are deeply concerned about the reliability of human performance.

The three performance levels can best be introduced by relating them to a familiar activity like controlling a vehicle on the road. For an experienced driver, the control of speed and direction occur almost entirely at the skill-based level. It would be largely impossible for them to tell you how precisely they changed gear or steered the vehicle. Maybe they could spell out the main steps involved, but they would not be able to explain in words how exactly they manipulated the controls. Most of the problems associated with driving relate to other road users. How we should behave in regard to the vehicles around us is clearly specified by the Highway Code and

Table 3.2 The principal characteristics of the skill-based, rule-based and knowledge-based levels of performance

Level	Features
Skill-based (SB)	Automatic control of routine tasks with occasional checks on progress.
Rule-based (RB)	Pattern-matching prepared rules or solutions to trained-for problems.
Knowledge-based (KB)	Conscious, slow, effortful attempts to solve new problems 'on line'.

road traffic laws. Thus, most things relating to the social aspects of road use occur at the rule-based level of performance. Almost all situations are covered by rules of the kind: if (situation X occurs) do—or don't do—(action Y). Sometimes, however, we have to give our minds over to dealing with an unplanned-for problem. We might, for example, be travelling at high speed along a main highway and we hear on the radio that there is a traffic snarl-up ahead of us. If we continued along our intended route, we would surely encounter long delays. So we have to find an alternative. Many people under these circumstances would not stop to consult a map. Instead, they would keep on driving and try to work things out in their heads. This would involve using an incomplete and probably inaccurate 'mental model' of the road geography ahead. Such strenuous and often erroneous 'mind-work' would be occurring at the knowledge-based level.

It should be stressed that these three levels of performance are not mutually exclusive. As the driving example shows, they could all three be operating at the same time. Thus, the driver whose way ahead is blocked could be working out an alternative route (KB level) while, at the same time, controlling the speed and direction of the vehicle (SB level) and also negotiating traffic signals and manoeuvring around other road users (RB level).

The driver example is useful, but there is also a more systematic way of defining the three performance levels. We have done most of the preparation for this in the preceding section. Figure 3.5 sets the scene.

In this rather busy picture, we have used the continua illustrated in Figures 3.3 and 3.4 to define a space—we will call it the 'activity space'. On to this space, we can map the three levels of performance.

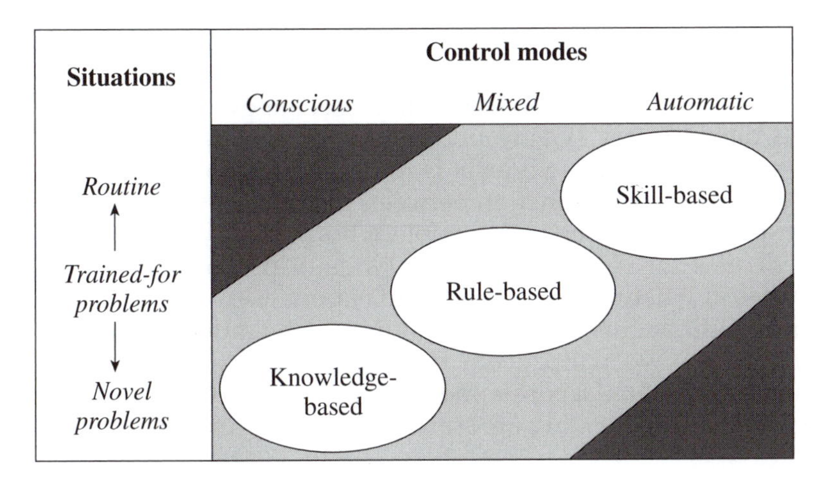

Figure 3.5 Using the activity space to define the three performance levels

At the top right-hand corner, there is the skill-based (SB) level in which we deal with familiar and non-problematic tasks in a largely automatic fashion. In the middle, there is the rule-based (RB) level when we modify our largely automatic behaviour because we have become aware of some change or problem. But this is a problem that we have been trained to handle, or have experienced before, or we have procedures that tell us what to do. In any event, we apply stored rules of the kind: *if X (some situation) then do Y (some action).* In applying these stored rules, we operate very largely by automatic pattern matching. That is, we unconsciously match the available signs and indications to some stored solution. Only after this—and sometimes not at all—do we use conscious thought to verify that we have adopted the right solution. At the bottom left-hand corner, we have the knowledge-based (KB) level in which we recognize that we have run out of ready-made solutions and have to think one out there and then. This is a slow and very error-productive business.

At the top left- and bottom right-hand corners (darker triangles), we have two kinds of pathological or maladaptive behaviour. In the former, we apply conscious thought to highly routine activities and so disrupt them. In the latter, we apply automatic responses to situations that require careful thought. This is what we do when we panic: we resort to primitive over-learned responses. In short, we 'run in circles, scream and shout'.

Stages in Acquiring a Skill

It is clear from what has been discussed above that performance levels depend very largely upon the degree of skill and training a person possesses. The SB level is the highest achievement of skilled performance. The distinction between the SB and KB levels depends upon expertise. Experts are people who possess a large number of stored rules for solving particular kinds of problems. The KB level switches in when we have run out of expertise.

Since skill acquisition is so important for understanding human performance, we will devote this short section to identifying the various stages that separate the novice from the virtuoso. The main stages in skill acquisition are summarized in Figure 3.6.

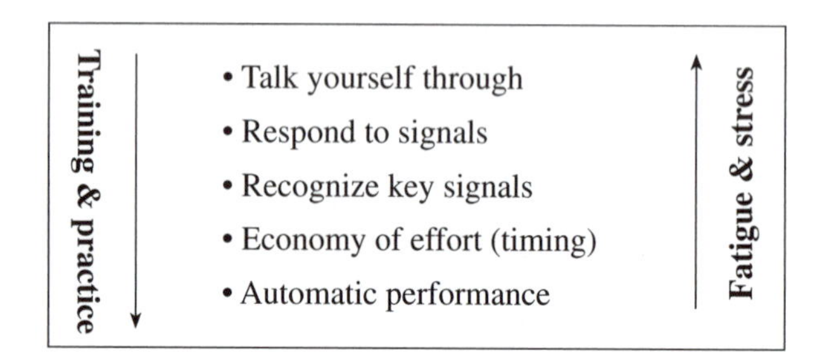

Figure 3.6 The principal stages in skill acquisition

When we start to learn a new skill, we rely heavily on words—either we talk ourselves through or we listen to the instructor. This is the verbal-motor stage. Next, we learn to connect our actions to various task-related signals (perceptual-motor stage). At first, all of these signals seem equally important, but then we become aware of the redundancies contained in them. We realize that we do not have to respond to all of them. When that happens, we no longer have to work so hard. We can space out our actions. In short, we start to possess the hallmark of the skilled person: the appearance of having all the time in the world. Finally, after much practice, we are able to perform the skill without being consciously aware of the individual actions. Another way of describing this transition from novice to expert is as a shift from the predominantly *feedback control* (largely employing the conscious workspace) to one involving long bursts of *feedforward control* (delivered from long-term memory).

Interestingly, fatigue and stress bring about the reverse of the acquisition process. The first thing to go is timing, then economy of action. Tired people work much harder. Unfortunately, they do not always realize what is happening and assume that their performance is just as good as ever.

Fatigue

'Fatigue' is a little word covering a wide range of effects, not all of which go together. They include feelings of tiredness, physiological changes and changes in performance relating to length of work and time of day. Time-of-day effects are relatively straightforward, so we will focus upon them. Figure 3.7 summarizes the main changes associated with the day–night (circadian) rhythms.

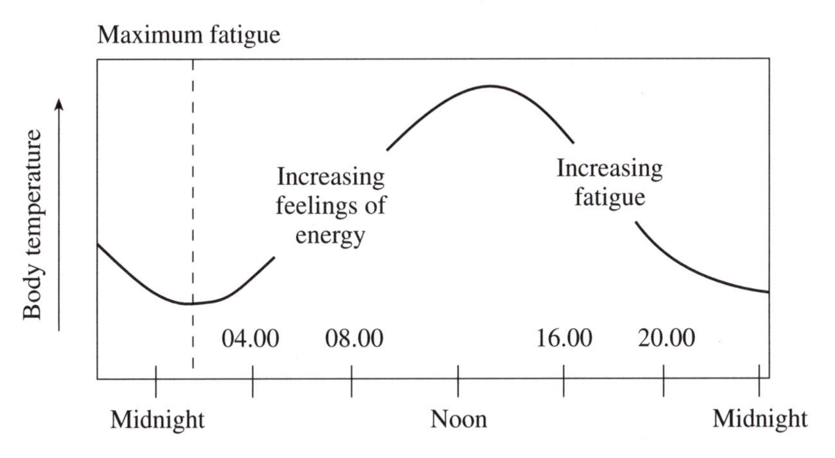

Figure 3.7 Time-of-day effects and fatigue

Body temperature fluctuates throughout the 24-hour cycle. The lowest point occurs at around 03.00 hrs, the highest at 14.00 hrs—the difference is about 0.5 degrees Celsius. Feelings of fatigue reach their peak in the early hours of the morning. These diminish and more energetic feelings increase until just past noon, whereupon they decline as they approach the low-point during the early hours of the morning. Such fluctuations are examples of circadian rhythms, *circadian* meaning 'around a day'. Performance differences correspond closely to the circadian temperature cycle. It is no coincidence that the errors contributing to several accidents and near-catastrophes occurred in the early hours of the morning (Three Mile Island, Bhopal, Chernobyl, the BAC 1–11 window blow-out, the loss of engine oil in a B737, the A320 spoiler incident and so on). There are also fairly wide individual differences in the circadian rhythm, with some individuals doing their best work in the morning and others in the afternoon and evening.

Circadian rhythms can be disturbed by external factors such as time-zone travel—jet lag—and shift changes. On average, it takes approximately one day to recover from each time zone travelled away from home. In general, recovery from east–west travel (phase delay in the sleep/wake cycle) is quicker than from west–east (phase advance), though recovery from a homeward journey is often faster than from the outward journey. Similarly, adjustments for changes from the day to the night shift (phase delay) are faster than from changes to an early morning shift (phase advance). Regular shift patterns alter body rhythms, while rotating shifts do not. However, rotating shifts mean working at the lowest point in the efficiency cycle and the highest point in the fatigue cycle. More will be said about fatigue and shift work in later chapters.

Stressors

'Stress' is another small word for a wide range of complex effects. The various types of stressors that are likely to be encountered in maintenance work and could have an adverse effect upon performance are summarised below.

- *Physical stressors*. Heat, humidity, confined spaces, noise, vibration and the like.
- *Social stressors*. Anxiety, group pressures, incentive schemes, disciplinary action and so on.
- *Drugs*. Alcohol, nicotine, medication and so on.
- *The pace of work*. Boredom, fatigue, interruptions and time pressures.
- *Personal factors*. Domestic worries, aches and pains, colds and generally feeling below par.

Stresses resulting from significant life events such as divorce, financial worries and the like can reduce general well-being and increase the susceptibility to some illnesses. The effects of these troubles can spill over into the workplace. People who are experiencing such events may be distracted by intrusive thoughts, particularly when workload is low. In addition, people can take more risks when they are under emotional stress brought on by life events such as marriage problems or financial problems. For example, it has been found that people who are facing high levels of life stress are less likely to wear a seat belt when driving.[3]

Arousal

Arousal is the body's reaction to stresses, biological drives and motivational influences. It ranges from very low arousal (sleep, torpor, drowsiness) to very high arousal (agitation, strong emotion, panic). Too little or too much arousal is associated with less than optimum performance, as illustrated in Figure 3.8—the *inverted U-curve.*

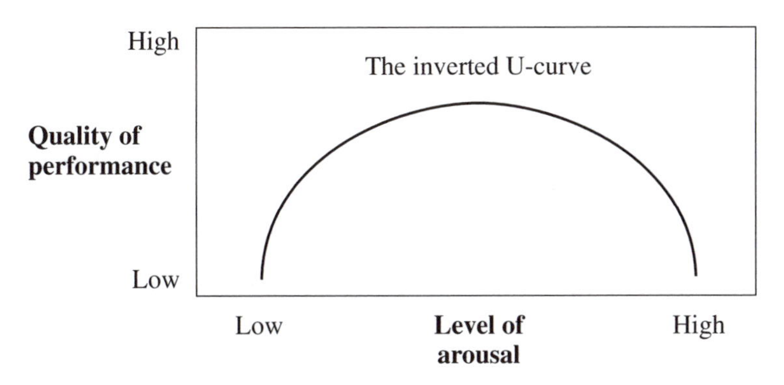

Figure 3.8 The relationship between arousal and the quality of performance

Arousal has important effects on attention. At low levels of arousal, our attention tends to be poorly focused, and can drift easily to task-irrelevant information, as we seek distraction from our boredom. At high levels of arousal, on the other hand, attention becomes highly focused, causing people to neglect or miss peripheral but task-relevant cues. Once again, there are individual differences. Extraverts, for example, tend to be more relaxed than introverts. As a result, extraverts may work better under pressure than introverts, but introverts may be better than extraverts at keeping their attention on boring jobs.

Coping with Informational Overload

One of the important differences between machines and human beings is that machines tend to break down abruptly when overloaded, while people degrade gracefully. This is especially the case when human beings are faced with increasing informational demands. They use a variety of strategies to shed informational load. Listed below is a progression of steps that people employ in coping with informational overload.

- Ignore selected inputs
- Trade-off accuracy for speed (the speed–error trade-off)
- Postpone things until quieter times
- Reduce the level of discrimination—accept coarser matches
- Redistribute the work, if possible
- Abandon the task altogether.

Personality Types

People differ not only in their abilities but also in their basic personality types. Different personality types are associated with characteristic performance styles. From ancient times, this has been seen as having a physiological or bodily basis. For example, the 'Doctrine of the Humours' classified people according to which of the four bodily humours or fluids was predominant. Those with an excess of blood were described as *sanguine*, those with an excess of phlegm were *phlegmatic*, those with an excess of black bile were called *melancholic*, and those with an excess of yellow bile were termed *choleric*. These terms have survived into contemporary language.

Modern psychologists have devised a comparable classification of personality types. This is formed on the basis of two independent dimensions: *introversion–extraversion* (based on the degree to which the nervous system seeks external stimulation), and *stable–unstable* (based upon arousal levels). Together, they make up a four-part categorization that corresponds very closely with the Doctrine of the Humours. The characteristics of these four types are summarized in Figure 3.9.

Unstable introverts correspond to the melancholic type, unstable extraverts to the choleric type, stable introverts to the phlegmatic type and stable extraverts to the sanguine type. Although the notion of accident proneness (the idea that certain people are consistently more liable to mishaps than others) is no longer widely accepted, there is some evidence to suggest that unstable extraverts are more likely to be involved in accidents than are other personality types. Different personality types are suited to different professions. Phlegmatic individuals (stable introverts) tend to make good anaesthetists and air traffic controllers, while sanguine people tend to become good surgeons or pilots. But these distinctions are by no means hard and fast. Like intelligence, personality types tend to follow a normal or bell-shaped frequency distribution with the great majority of people occupying intermediate positions along both the introversion–extraversion and stable–unstable dimensions, with relatively few at the extremes.

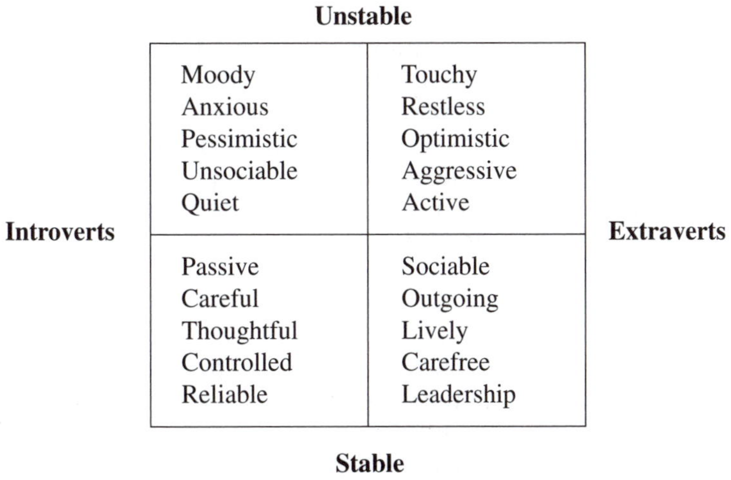

	Unstable		
Introverts	Moody Anxious Pessimistic Unsociable Quiet	Touchy Restless Optimistic Aggressive Active	Extraverts
	Passive Careful Thoughtful Controlled Reliable	Sociable Outgoing Lively Carefree Leadership	
	Stable		

Figure 3.9 Four personality types based upon the introversion–extraversion and stable–unstable dimensions (after Eysenck)

Biases in Thinking and Decision Making

Ideally, we would make all our work-related decisions by following a careful rational process in which we consider all options, evaluate each in turn and then select the best course of action. However, in reality things are somewhat different. We have learned that workplace decisions, including those made by maintenance personnel can be biased and distorted by a range of factors. Sometimes we do not consider all the alternatives, or we take labour-saving mental short cuts to arrive at a solution. Or under emotional or time pressures, our thinking is 'short circuited' so we end up deciding on a course of action which is unsuitable for the situation we are in. Two common decision making problems are outlined below.

Confirmation Bias

A powerful factor that can interfere with thinking is the tendency known as 'confirmation bias'. When faced with the need to solve an ambiguous or ill-defined problem, we frequently develop a theory to explain the situation. Once we have such an idea, we tend to search for information that will *confirm* what we suspect. People rarely attempt to prove themselves wrong, and in fact sometimes disregard information that would contradict their ideas. In troubleshooting a fault, for example, a maintainer may decide on the most likely cause, and then attempt to prove that this is the problem. The initial incorrect fault diagnosis can block attempts to consider other possibilities. For instance, an aircraft was reported to be pulling to the left when brakes were applied. The maintenance personnel considered that this was because the brakes were binding on one of the left wheels. After time-consuming and unsuccessful attempts to correct the problem, a quite different problem was eventually discovered. The brakes on the right-hand side were not working.

Emotion and Decision Making

There is a clear link between frustration and aggression. It seems to be hard-wired into people that if a situation keeps frustrating us, we may move into 'aggressive mode', giving us access to extra reserves of strength and motivation. Thousands of years ago, the frustration–aggression link probably helped our ancestors fight wild animals, catch food and solve many other Stone Age problems. Unfortunately, the world in which we live requires careful action more often than brute force, but we still carry within us this primitive problem-solving system. Frustrating circumstances in maintenance such as a lack of proper tools, delays and unreasonable pressures can all combine to cloud judgement with emotion.

Summary

This chapter has sought to introduce maintainers and other technically minded, non-psychologists to some of the factors that control and influence human performance. We began by identifying the contrasting properties of two complementary and interacting mental structures: the conscious workspace and long-term memory. These, in turn, provide two distinct kinds of action control: the conscious mode and the automatic mode. Three performance levels were identified: the skill-based, rule-based and knowledge-based levels. Which of these levels is dominant at any one moment depends upon both the control mode and the nature of the situation. The three performance levels are heavily dependent upon a person's degree of skill. They will form the basis of the error classification to be described in the next chapter.

The remainder of the chapter examined a number of different influences upon human performance. These included the 24-hour bodily rhythm and its associated feelings of fatigue and energy, various factors likely to induce stress, the notion of arousal and the inverted U-shaped relationship with performance quality, the methods people use to cope with information overload, personality differences and, finally, biases in decision making. We are now equipped to concentrate upon the main purpose of this book, the nature, variety and management of human errors as they occur in maintenance-related activities.

Notes

1 J. Reason, *Human Error* (New York: Cambridge University Press, 1990).
2 J. Rasmussen, 'Human errors: a taxonomy for describing human malfunction in industrial installations', *Journal of Occupational Accidents*, **4**, 1982, pp. 311–35.
3 A. Schichor, A. Beck, B. Bernstein and B. Crabtree, 'Seat belt use and stress in adolescents', *Adolescence*, **25**, 1990, pp. 773–9.

4 The Varieties of Error

What is an Error?

Everyone knows what an error is—except perhaps psychologists whose business it is to come up with an agreed definition. This they have not done. Indeed, there are some psychologists who would deny the existence of errors altogether. We will not pursue that doubtful line of argument here. Instead, we will begin by stating a working definition that has shown itself to be useful in real-world settings.[1]

> An error is the failure of planned actions to achieve their desired goal, where this occurs without some unforeseeable or chance intervention.

The rider is important because it distinguishes controllable or voluntary actions from those shaped by mere luck, either bad or good. If, for example, you were suddenly struck down by a chunk of returning space debris, you would probably not arrive at your intended destination, but you could hardly be said to be in error. Conversely, someone who slices a golf ball so that it hits a passing bird and rolls on to the green may achieve his or her purpose, but the actions are still erroneous.

All errors involve some kind of deviation—the departure of actions from their intended course, or the departure of planned actions from an adequate path towards some desired goal, or the deviation of work behaviour from appropriate operating procedures. Sometimes, these deviations involve rule violations, such as driving over the speed limit. For the purposes of this book, violations will be treated as a separate category, even though they can be committed erroneously.

In essence, there are three ways in which planned actions may fail to achieve their current goals.

1 The plan of action may be entirely appropriate, but the actions themselves do not go as planned. These error types are called

skill-based errors and include *slips, lapses, trips* and *fumbles* (*slips and lapses* for short). Here, the failure involves either attention (*slips of action*) or memory (*lapses*).

2 The actions may go entirely as planned, but the plan is inadequate to achieve the desired goal. These error types are called *mistakes*. They can be split into two classes: *rule-based mistakes* and *knowledge-based* mistakes. Whereas slips and lapses occur at the level of execution, mistakes arise when dealing with a problem (a departure from the expected course of things) and involve failures at the level of formulating an intention or a making a plan.

3 The actions can deviate intentionally from the safe method of working. Such violations may involve a contravention of formal rules and procedures, but they can also deviate from unwritten norms or standard practice.

Listed below are the major types of unsafe act that occur in maintenance. We will consider each of these in the remainder of this chapter.

- Recognition failures
- Memory lapses
- Slips of action
- Errors of habit
- Mistaken assumptions
- Knowledge-based errors
- Violations.

Skill-based Recognition Failures, Slips and Lapses

Skill-based errors can be identified with three related aspects of human information processing: recognition, memory and attention. We will consider these varieties of skill-based errors below.

Recognition Failures

Recognition failures fall into two main groups

- The *misidentification* of objects, messages, signals and the like.
- The *non-detection* of problem states (inspection or monitoring failures).

The principal factors contributing to misidentifications are listed below.

- *Similarity*. In appearance, location and function between the right and wrong objects.
- *Indistinctness*. Poor illumination and signal-to-noise ratios.
- *Expectation*. We tend to see what we want to see.
- *Familiarity*. In well-practised and habitual tasks, perceptions become coarser.

Misidentifications involve putting the wrong mental interpretation upon the evidence gathered by our senses. These errors have been the cause of many serious accidents. They include train drivers who misread a signal aspect and pilots who misinterpret the height information provided by their instruments (particularly the old 'killer' 3-pointer altimeter).

A major factor in misidentifications is the similarity (in appearance, location, function and so on) between the right and wrong objects. This can be made worse by poor signal-to-noise ratios (that is, poor illumination, inaccessibility and the like). For example, an aircraft maintenance worker needed to top up the hydraulic fluid on an aircraft. Only after he had added the fluid did he realize that he was holding a (now empty) tin of engine oil. Oil and hydraulic fluid were stored in nearly identical tins in a poorly lit storeroom.

Misidentifications are also strongly influenced by expectation: we tend to see what we expect to see. What we perceive is derived from two types of information: the evidence of our senses and knowledge structures stored in long-term memory. The weaker or more ambiguous the sensory evidence, the more likely it is that our perceptions will be dominated by expectation, or our stored knowledge structures. As outlined in Chapter 3, once we have formed an idea about what is going on, we tend to select information that will confirm this hunch, even when there is contradictory evidence available. Strong habits are also like expectations: we sometimes accept a crude perceptual match to what is familiar or anticipated, even when it is wrong.

Despite new techniques and technology for detecting faults, we still rely on the human eyeball for most fault-finding tasks. Non-detection errors typically involve a failure to notice a visible fault during an inspection.

Some of the main factors contributing to non-detection errors are summarized below.

- Inspection was interrupted before reaching defect.
- Inspection was completed, but the person was distracted, preoccupied, tired or in a hurry.
- The person did not expect to find a problem in that location.
- One defect spotted, but next one close to it missed.

- Inadequate lighting, dirt, or grease.
- Inadequate rest breaks.
- Access to job was unsatisfactory.

Other factors include inexperience and not being sufficiently trained in knowing what to look out for in the way of signs and symptoms. At other times we do not take into account the physiological limitations of the human visual system. For example, some specialist inspection techniques are carried out under low-light conditions, yet inspectors may be reluctant to wait the 10 minutes or more it takes for eyes to adapt to the darkness. Colin Drury and his colleagues have developed excellent guidance material on the human factors of maintenance inspections.[2] Non-detection errors also reflect the vigilance decrement described in Chapter 3. On a long, boring inspection task, the mind will tend to wander to other matters. It is no surprise that problems are often detected when they are not being looked for, as illustrated by the following example.

> After being on duty for 18 hours on a long overtime shift, the maintainer was carrying out a general inspection on an engine at around 22.00 hrs. He missed obvious damage to the internals of the cold stream duct area. The damage was only found after investigation of another defect.

Recognition and non-detection errors: a case study

In 1990, a flight deck windscreen on a BAC 1-11 blew out as the aircraft climbed through 17 000 feet over Oxfordshire, UK.[3] The captain was sucked out and remained half in and half out until the aircraft landed at Southampton. He survived with relatively minor injuries, though it was a close run thing for both the passengers and the crew.

A scheduled windscreen change had been carried out on the previous night. The accident was due in large part to two recognition errors—a misidentification and a non-detection—made by the shift supervisor who, being short-handed, elected to do the job himself. The bolts he removed from the old windscreen were 7Ds, but many were dirty and he believed he needed around 80 new bolts. He went to the store man to get them, but discovered that there were only a handful of 7Ds in stock. So he drove off to the ramp and searched through an unsupervised and poorly labelled carousel containing general aviation fixings. He found what he thought was a drawer full of 7D bolts and sought to match them with one he had taken from the old windscreen. The matching technique he used was to hold the used bolt and a new bolt between finger and thumb and inspect them, but the area was poorly lit and he was not wearing his reading glasses. In the event, he wrongly selected thinner 8C bolts.

Second, he failed to detect his error when he was fitting the windscreen. There were a number of contributing factors. Due to a poorly positioned platform, he had to lean across the front of the aircraft. The magnetic bit-holder of his torque screwdriver was ineffective so he had to hold the bit in with his left hand while fitting the bolts. This meant that his hand concealed the larger amount of countersink that should have told him that he was using thinner bolts. The bolts fitted into elliptical nuts, so he did not notice the different torque feel.

This case study illustrates an important fact about maintenance errors: they often occur in a linked sequence, with one error leading on to the next. Making an error can greatly increase the chances of making a subsequent one. This has been called an *error cascade*. Error cascades can also involve a number of different people each making isolated errors that link together to punch a dangerous hole in the system's defences. We will see examples of these cascades in Chapter 6 when we come to consider three fatal maintenance-related accidents in detail.

Memory Failures

A survey of Australian aircraft maintenance personnel collected over 600 reports of maintenance incidents. Memory lapses were the most common form of error, being implicated in 20 per cent of incidents.[4]

Memory failures can occur at one or more of three information-processing stages.

- *Encoding (or input) failure*. Insufficient attention is given to the to-be-remembered item so that it is lost from short-term memory (the conscious workspace).
- *Storage failure*. Remembered material decays or suffers interference in long-term memory.
- *Retrieval (or output) failure*. Items that we know we know cannot be recalled at the required moment.

Input failures

Common *input failures* include failing to remember something we are told and not taking note of previous actions. What are we most likely to forget on being introduced to someone? Their name. Why? The name is part of a flood of new information about this person and often fails to get taken in unless we make a special effort to focus on it (then we often cannot remember what they looked like or what they did for a living). This tells us that giving just the right amount of attention to something is an important precondition for being able to remember it later.

The second kind of *input failure* is the 'forgetting' of previous actions. Actually, not remembering is a better description. Again, this is due to a failure of attention. When we are carrying out very familiar and routine tasks, our minds are almost always on something other than the job in hand. This, as explained in Chapter 3, is necessary in order for the task to be done smoothly. The result is that we 'forget' where we put our tools down, or find ourselves walking around looking for something that we are still carrying.

Other varieties of *input* or *encoding failures* include the following:

- *Losing our place in a series of actions*. We 'wake up' in the middle of a highly routine task and don't immediately know where we are in the sequence. Place-losing errors have a potentially dangerous sequel: getting it wrong when trying to find your place again. Two kinds of place-finding errors can occur: either we judge ourselves to be further along than we actually are in the sequence, and so omit some prior step; or, less dangerously, we judge ourselves to be not as far along as we actually are and hence repeat a step unnecessarily (for example, putting double the amount of tea in the teapot).
- *The time-gap experience*. We can't remember things about where we have been walking or driving in the last few minutes, or what exactly we have been doing. For example, we can be in the shower and cannot remember whether or not we have put shampoo on our hair. The evidence (if there was any) has been washed away, and we have been thinking about something else. In short, we have not been attending to the routine details.

Storage failures

These come in a number of forms, but the one most likely to have an adverse effect on maintenance-related activities concerns forgetting the intention to do something. An intention to perform some act or task is rarely put into action immediately. Usually, it has to be held in memory until the right time and place for its execution. Memory for intentions is called *prospective memory*, and it is particularly prone to forgetting or sidetracking, so that the action is not carried out as intended. Many maintenance personnel are familiar with the sudden thought on the drive home of 'Did I or didn't I?' Did I replace that fuel cap? Did I remove that tool?

It is, of course, possible to forget an intention completely, so that no trace of it remains. More usually, the forgetting occurs in degrees. Almost forgetting the intention completely can reveal itself as the 'I should be doing something' feeling. Here, you have a vague and uneasy sense that you should be doing something, but cannot remember what or where it should be done. Another fairly common experience is that you remember the intention and start to carry it through, but somewhere along the line—usually because you are preoccupied with or distracted by something else—you forget what it is that you came to do. The place could be a shop, a parts store, or you could find yourself standing in front of an open drawer or cupboard at home. You simply cannot recall what it is you came to fetch. This is the 'What am I doing here?' feeling. The third possibility is that you set out to perform a plan of action, think you have completed it, but later discover that you have left something out. A common experience is to return home to find a letter you intended to post still lying on the hall table.

Retrieval failures

These are among the commonest ways that your memory can let you down. And, as you may have noticed, they become increasingly common with age—the 'What's his name?' experience.

Retrieval failures can show themselves as *tip-of-the-tongue* (TOT) states when you realize that you cannot call to mind a name or a word that you know you know. The searched-for word seems tantalizingly close—on the tip of your tongue, in fact. The problem is usually made worse because some other word or name comes into your mind, but you know it is not the one you are trying to find. However, you also have a strong sense that somehow it is close to the target item. You may feel it sounds similar, or has the same number of syllables, or is a name that belongs to someone who is related to or works with the person whose name you are trying to find.

Omissions following interruptions

A failure to carry out a necessary check on progress can be caused by some local distraction. For example, you intend to collect a manual, but on removing it from the shelf other books fall down. You replace the books but depart without the manual you came to get.

On some occasions, the interruption causes the person to 'forget' the subsequent actions, or allows him or her to get sidetracked into something else. On others, the actions involved in dealing with an interruption get unconsciously counted in as part of the original action sequence. For example, you are making tea and find that the tea bag packet is empty. You go to the cupboard and open up a new packet. Then you pour boiling water into an empty teapot, having omitted to put in the tea bags.

Premature exits

As the name implies, premature exits involve terminating a job before all the actions are complete—like getting into the shower with your socks on. As we approach the end of a routine task, our minds jump ahead to the next activity, and may lead us to leave out some late step in the first task. For example, there have been cases where compressed air was used to test hospital oxygen distribution systems, but at the completion of the test, the maintainers left the system connected to air rather than oxygen.

Skill-based Slips

In Chapter 3 we described how in familiar situations, much of our behaviour is guided by 'automatic' routines. The more skilled and experienced a maintainer is, the more he or she will be able to perform quite complex tasks on 'mental autopilot'. In everyday life we can change gears in our car, tie shoelaces, comb our hair, or answer the phone with very little conscious attention. Maintenance work contains many familiar situations where automated skill routines develop, and without these pre-packaged routines work would progress at a snail's pace. Closing access covers, replenishing fluids, pulling circuit breakers, securing screws and checking pressures are each examples of maintenance tasks that, because they are relatively predictable and routine, can be delegated unconsciously to skill-based action routines.

People do not necessarily choose to perform tasks in this way. Whether we like it or not, automated skills can start to take over our actions in familiar situations. Sometimes we only become aware of an automatic programme running in the background when it causes us to do something we never intended. Writing in 1890, the pioneering American psychologist William James noted that if asked how they performed a familiar action, the response of most people would be 'I cannot tell the answer, yet my hand never makes a mistake'.[5]

A good test to establish whether a task has reached this automated level is to consider whether the person can hold a conversation while they perform it. An inexperienced driver will stop talking when approaching a corner, or negotiating heavy traffic. Yet once we have developed skill routines to deal with these situations, our attention can be directed elsewhere, enabling us to hold a conversation, listen to the radio or plan dinner, while still controlling the car.

Action slips happen when our automated routines take control of our actions in ways that we never intended. For example, an electrician had been asked to change a light bulb that indicated whether a hydraulic on/off switch was selected. The hydraulic system was being worked on and the electrician was aware that it would be unsafe to activate the system. Nevertheless, after changing the bulb, and before he had realized what he was doing, he had followed his usual routine and pushed the switch to the 'on' position to test whether the light was now working.

Our skill routines are like hijackers lying in wait to take over our actions. Skill-based slips can be particularly dangerous because we can find our unconscious mind making us do things which our conscious minds would never consider doing.

Figure 4.1 shows the main features involved in a typical action slip. Imagine that you are carrying out a highly practised action routine, like boiling an electric kettle preparatory to making a beverage. Imagine also that you have a guest who has asked for tea, while you are a habitual coffee-drinker. You go to the kitchen, fill the kettle and set it to boil. In the meantime, you start thinking about some important current concern. As a result, you miss the choice point and fill both cups with instant coffee and pour on the water. In this case, the kettle sequence is the fat arrow on the left. The fatter of the two arrows on the right is the coffee-making routine. The thinner arrow is the tea-making routine. You miss the choice point and your actions run, as on rails, along the familiar route. But this time, because of a change in circumstances, it is an absent-minded slip.

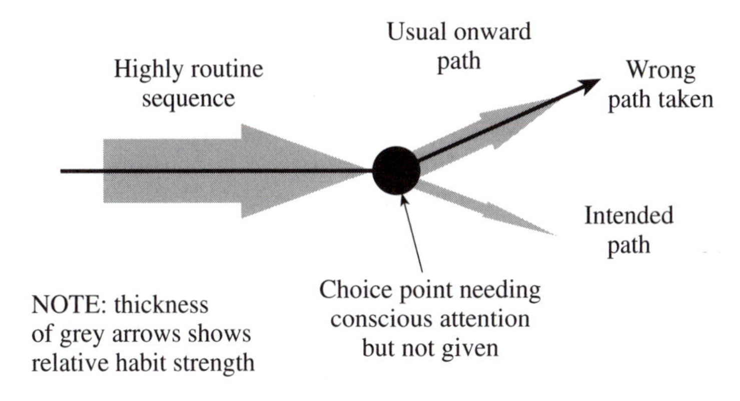

Figure 4.1 The typical pattern of events involved in an action slip

Everyday examples of such action slips include driving to work on a Saturday morning when you meant to go to the shops. Or, you intend to stop off to buy groceries on the way home, but go straight past. The former is a *branching error*, the latter an *overshoot error*.

Branching slips, as the name indicates, involve actions where two different outcomes have an initial common pathway. In the beverage-making example given earlier, boiling a kettle of water is the first stage in achieving a variety of goals: making tea, making coffee, speeding up the cooking of vegetables and so on. The defining feature of these slips is that the wrong route (that is, the one not currently intended) is taken. This 'wrong route' is almost invariably more familiar and more frequently travelled than the one that was currently intended. The slip is then triggered by a change in the normal routine (as in the tea-making slip described earlier). If there had been no change, there would probably have been no error, as the actions would have run on automatically to their normal conclusion irrespective of being distracted or preoccupied. *Overshoots* operate in a similar way, except that some intended deviation from the normal pattern is omitted.

Factors Promoting Action Slips

By now it will be appreciated that absent-minded slips of action are not random events. They fall into predictable patterns and are associated with three distinct causal factors.

1 The performance of a routine, habitual task in familiar circumstances.
2 Attention is 'captured' by some unrelated preoccupation or distraction.
3 There is some change, either in the plan of action or in the surroundings.

Paradoxically, absent-mindedness is the penalty we pay for being skilled; that is, for being able to control our routine actions in a largely automatic fashion. It is therefore natural that slips and lapses are most likely to occur during the execution of well-practised habitual tasks in familiar surroundings. Of course, we do commit errors when we are learning a new skill (like using a computer keyboard), but these errors are most likely to be fumbles and mis-hits due to inexperience and lack of motor coordination.

Attention is a limited commodity. If it is given to one thing it is necessarily withdrawn from other things. 'Capture' happens when almost all of this limited attentional resource is devoted to something unrelated to the task in hand. If it is an internal worry, we call it preoccupation; if it is something happening externally in our immediate vicinity, we call it distraction. Attentional capture, of one kind or another, is an indispensable condition for an absent-minded slip or lapse. Attention to some critical choice point in the task is the thing that goes absent in absent-mindedness.

Many action slips involve carrying out actions that are commonplace or habitual in that particular situation—we call these *strong-*

but-wrong slips. The wrong actions are often perfectly coherent; they are simply not what was intended at that time. The trigger for the slip is some kind of change, either in the intention or in the local circumstances. If that change had not occurred then the actions would have run along their accustomed tracks as intended. Thus, change of any kind is a powerful error-producer.

Rule-based Mistakes

For the most part, maintainers are extensively trained and their work is highly proceduralized. This means that most of their mistakes—failures at the level of formulating intentions or problem solving—need to be understood as a deviation from the appropriate rule or procedure. Rules can exist within maintainers' heads, put there through training and experience, or they can be written down in manuals and standard operating procedures.

There are two main ways in which rule-based errors can arise in maintenance-related activities.

- *Misapplying a good rule (assumptions).* That is, a normally good rule can be used in a situation for which it is not appropriate, perhaps because of habit or a failure to detect a change in the circumstances.
- *Applying a bad rule (habits).* A bad rule is one that may get the job done on this occasion but can have unwanted consequences.

In many respects, violations—failing to apply the good rules—represent a distinct class of unsafe acts, having a number of important differences from errors. In what follows, we will treat the misapplication of good rules and using bad rules as simple rule-based mistakes. Violations, on the other hand, are sufficiently important in their own right to be dealt with separately, even though they may—and often do—involve mistakes.

Misapplying Good Rules (or Assumptions)

A 'good rule or principle' is one that has proved its worth in the past. The rule could be written down, or it may exist as some simple 'rule of thumb'. Misapplying good rules can happen in circumstances that share many common features for which the rule was intended, but where significant differences are overlooked. For example, a maintainer may have developed the 'rule' that the correct tyre pressure is such-and-such pounds per square inch. An assumption like this works well so long as nothing changes, but if an exception to the rule is encountered, the rule will lead to an error. The business of applying problem-solving rules is often complicated by the fact that different problems can share common elements. In other words, it is possible that a given problem presents both indications suggesting that the common rule (common because it is a useful rule) should be applied and counter-indications directing the person to apply a less commonly used rule.

> A maintenance crew had commenced an A-check on a twin-engine jet aircraft. There was a requirement to lock out the thrust reverser on an engine, although there was no work to be done on the thrust reverse system. The crew deactivated the reverser; however, there was no requirement to write up the deactivation in the defect log, and hence no log entry was made. After a shift change, a second crew completed the A-check. The task card called for system reactivation; however, because no work had been done on the reverser system, the crew did not expect the reverse thrust to have been deactivated and did not check the status of the lockout plate. The aircraft was dispatched with an inoperative and undocumented reverse thrust system.

In the case study above, the second shift made an unspoken assumption about how the first shift had prepared for work on the aircraft. Clearly, this assumption did not occur randomly, but was shaped by the documented procedures being used at the time. After this incident (and others like it) the airline changed its procedures to make sure that important task steps were specified more clearly.

Another maintenance error that involved an assumption occurred in 1994, when a Boeing 747-200 dropped and then dragged its Number 1 engine during the landing rollout at New Tokyo International Airport, Narita.[6] The immediate cause of this accident was the migration of a fuse pin from its fitting. This, in turn, was the result of a failure to replace the secondary fastenings of this fuse pin during a major check. Prior to this part of the check, an inspector had marked the work card steps covering the replacement of the secondary pin retainers as 'N/A' (not applicable). He was wrong. But the reason for his mistake was understandable. The Boeing Aircraft Company had recently requested that all B747s of this type be fitted with secondary fastenings after a run of fuse pin accidents across the world. At this point, however, only seven of the airline's fleet of 47 B747s had been modified. Thus, the inspector did not believe that secondary retainers were required on this aircraft, nor did he realize that these retainers had been removed. This was a rule-based error: the general 'rule-of-thumb' being that secondary retainers were not needed. But this was one of the few modified aircraft and so was an exception to this 'rule'.

Applying Bad Rules (Habits)

Many people pick up bad habits when learning a job. The trouble is that such bad habits become a part of a person's established work routines. No one corrects them. They get the job done. And most of the time there are no bad consequences, at least not until the person encounters circumstances that expose the flaws in their habit or rule. For example, before pressurizing hydraulics on any piece of equipment with hydraulically powered systems, whether it is earth moving equipment or an aircraft, a good rule is to ensure that controls have not been moved while the hydraulics were off. Otherwise unpleasant surprises can ensue when systems start moving unexpectedly the moment the hydraulics are activated. A maintainer can go about their job with the bad rule 'to start hydraulics, just press the on switch immediately' and not experience any incident to tell them that they are applying a bad habit or bad rule. But when, say, someone has moved the flap or gear selectors on a parked aircraft, this bad rule creates a serious hazard.

Perhaps the saddest example of a bad rule was the primary cause of the catastrophic rail collision at Clapham Junction in 1988 (see also Chapter 6).[7] A northbound commuter train ran into the back of a stationary train after having passed a green signal. Thirty-five people died and 500 were injured. But the signal had failed unsafe, a failure that was directly connected to the working habits of a technician engaged in rewiring the signal on the previous day. Rather than cutting off or tying back the old wires, the technician merely bent them back out of the way, or so he believed. It was also his practice to re-use old insulating tape (another bad rule), though on this occasion no tape at all was wrapped around the bare ends of the wire (a violation). As a result, the signal came into contact with nearby equipment causing a 'wrong-side' signal failure. The technician in question was a highly motivated and hardworking person who had never (in his 12 years of service) received any proper training. He had picked up the job by watching other people and trying things out for himself. As a result, his bad rules went uncorrected.

Knowledge-based Errors

The skill–rule–knowledge distinction was introduced in Chapter 3. You will recall that knowledge-based problem solving comes into play when we are faced with new problems or situations where we have to go back to 'first principles' to understand what actions to take. An analysis of the work of aircraft maintainers found that they spend less than 4 per cent of their time dealing with such situations.[8]

For example, Alan Hobbs was with an experienced maintainer who was performing a routine inspection of the cargo compartment of a B767. He came across a crystalline powder caked on to the floor of the compartment, partly covering some of the pallet rollers. Where had it come from? Did it pose a threat to the aircraft? Or did it indicate an airworthiness problem? After putting a small amount on his hand and smelling it, he concluded that it was food material that had been spilled. The area was cleaned and the inspection was signed off.

Although most situations calling for knowledge-based problem solving, like the example above, are resolved without incident, these tasks have the highest error rate of all the situations faced by maintenance personnel. The errors that occur during knowledge-based problem solving can arise for two reasons, either failed problem solving, and/or a lack of system knowledge. The following report illustrates an error that involved each of these issues.

> I wanted to turn the radio master on but could not find it, as the switches were poorly marked or unreadable. I was unfamiliar with the aircraft, so I asked an airframe maintenance engineer who was working on the aircraft and he pointed to a red rocker switch. I queried him and he said that must be it. I pushed the switch and the right engine turned over, with the propeller narrowly missing a tradesman who was inspecting the engine. There is no radio master in this aircraft. I immediately marked the 'start' and some other switches and learned a valuable lesson.

Knowledge-based errors are particularly likely if the person is performing a task for the first time. This is illustrated by the following example.

> A modification was called for on the brakes of a large twin-engine jet aircraft. The licensed maintainer, who was carrying out this job for the first time, misinterpreted the documentation and installed a component upside down. Another worker who had performed this modification previously noticed the fault and the error was rectified before the aircraft departed.

Although inexperienced maintainers are the most likely to make knowledge-based errors, even experienced personnel get caught out from time to time. New or unfamiliar tasks, unusual modifications or hard-to-diagnose faults can all breed knowledge-based errors. Nearly 60 per cent of maintenance personnel report that they have continued with an unfamiliar job, even though they were not sure they were doing it correctly (see Figure 4.2).[9] The saving grace is that people can generally recognize that they are faced with a novel problem and may be able to call on help, whether from technical support or from colleagues.

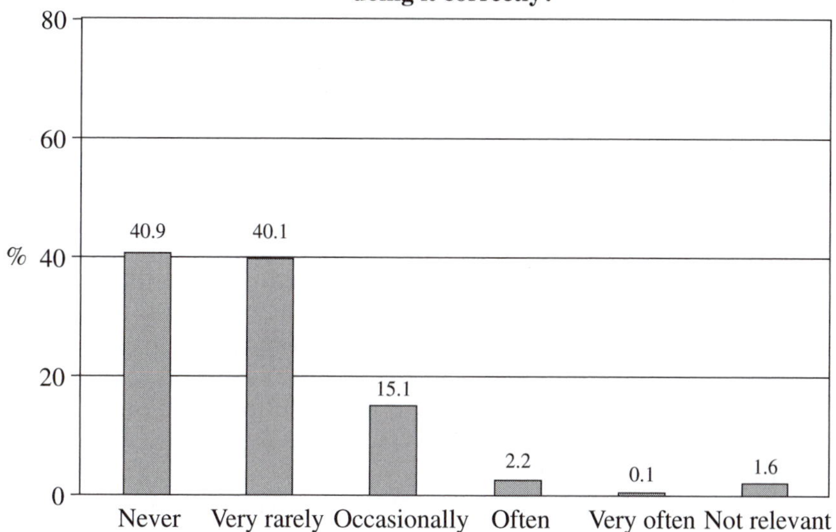

Figure 4.2 Aircraft maintenance personnel questioned on whether they have continued with an unfamiliar job despite being uncertain how to proceed

Source: A. Hobbs and W. Williamson, *Aircraft Maintenance Safety Survey – Results* (Canberra: Australian Transport Safety Bureau, 2000).

Violations

Up to this point, we have considered some of the classic varieties of human error. But there is an important type of dangerous act that is significantly different to the types of error discussed so far. These are violations. In recent years, psychologists and safety researchers have begun to discover just how widespread such actions are in safety-critical workplaces such as maintenance. In environments as diverse as oil production, medicine and nuclear power generation, we know that operators deviate from standard procedures and cut corners from time to time. Most maintenance is highly regulated, and personnel are expected to carry out their duties while observing legal requirements, manufacturers' maintenance manuals, company procedures and unwritten norms of safe behaviour. Yet such a well-intentioned proliferation of rules and procedures may serve to reduce the range of permitted actions to such an extent that a worker may find it difficult to complete the job *without* violating.

Although the distinction between errors and violations can be blurred in some cases—particularly when the violation is a mistake or the violator does not understand the consequences of non-compliance—there are a number of important differences between them. These are summarized below.

- *Intentionality*. Slips, lapses or mistakes are unintended. But people usually commit procedural violations quite deliberately (except where they are so ingrained as to have become automatic). It is important to note, however, that while people may intentionally carry out the non-compliant actions, they do not generally mean to bring about the occasional bad consequences. Only saboteurs intend both the non-compliant acts and their bad consequences.
- *Information versus motivation*. Errors arise from informational problems and are generally corrected by improving the information, either in the person's head or in the workplace. Violations, on the other hand, arise largely from motivational factors, from beliefs, attitudes and norms, and from the organizational culture at large. These are the things that need to be fixed if we are to reduce non-compliance to good rules.
- *Demographics*. Men violate more than women and the young violate more than the old. The same patterns do not occur for errors.

New research is uncovering the extent of violations. A study of European airlines by researchers from Trinity College Dublin revealed that 34 per cent of maintenance tasks had been performed in a manner that contravened the formal procedures.[10] In the Australian survey referred to earlier, 17 per cent of the reported maintenance occurrences involved violations.[11]

The prevalence of violations in aircraft maintenance may come as a surprise, even to airline managers. Common maintenance violations include:

- referring to unauthorized notes or 'black books' rather than approved sources
- deviating from formal documented procedures
- not making a system safe before working on it, perhaps because no one else seems to be around
- not performing a required functional check at the end of a maintenance procedure
- not using the correct tools for a job
- signing for checks which were not actually performed.

Violation Types

Like errors, violations take a number of different forms. There are three main categories.

- *Routine violations.* These are committed to avoid unnecessary effort, to get the job done quickly, to demonstrate skill, or to circumvent what seem to be unnecessarily laborious procedures.
- *Thrill-seeking or optimizing violations.* People have many goals and not all of them relate to the job. These violations are committed to appear macho, to avoid boredom or simply 'for kicks'.
- *Situational violations.* Sometimes it is impossible to get the job done if one sticks rigidly to the procedures. Here, the problem lies mainly with the procedure writers.

Routine Violations

The survey of Australian maintenance personnel revealed that routine violations were one of the most common forms of unsafe acts.[12] It was found that over 30 per cent of maintainers had signed-off a task before it was completed, over 90 per cent had done a job without the correct tools or equipment, and a similar number had not referred to the approved documentation on a familiar job. Those maintainers who confessed to performing more routine violations in their day-to-day work were also more likely to be involved in airworthiness incidents resulting in delayed aircraft, air turn-backs and the like.

There is a further penalty to these corner-cutting violations: they can form a normal part of maintainers' everyday working practices. In other words, they become established at the skill-based level of performance. The *principle of least effort* is a major force in human behaviour.

Thrill-seeking or Optimizing Violations

People, unlike robots, are complex creatures with a variety of needs. Some of these may relate to getting the job done efficiently, but others stem from more personal desires. The result is that the satisfaction of these job-related and personal needs become intertwined. A vehicle driver's need is to get from A to B, but in the process he or she (usually he) can seek to optimize the joy of speed and indulge aggressive instincts. Serious problems arise, however, when the gratification of these more basic instincts leads people to make errors. It is not necessarily the violations themselves that cause harm so much as the errors we make while violating safe operating procedures.

Driving at 110 mph is not, by itself, usually sufficient to cause a crash in a modern vehicle. But travelling at such a speed is likely to be a relatively unfamiliar experience. As such, we are more prone to misjudging the vehicle's handling characteristics. This has two bad consequences: not only are we are more likely to make an error; there is also a high probability that this error will have a bad outcome. Violations plus errors equal disaster.

These tendencies to optimize non-functional goals can become a part of an individual's personal style of working. This is particularly likely in the case of young males for whom testing the boundaries is a natural instinct. It is not by chance that men aged between 18 and 25 are at the greatest risk of road traffic accident fatalities. As mentioned earlier, these age and gender differences are not evident in error liability, at least during the span of a normal working lifetime.

Thrill-seeking or optimizing violations are not as common in maintenance as corner-cutting or routine violations, but they do occur from time to time. Practical jokes or initiation rites are a prevalent form of optimizing violation. Even towing aircraft can provide opportunities for optimizing violations as the following example illustrates.

> The aircraft was at the terminal and had to be towed to the run up bay. A maintainer was working in the engine on the variable inlet guide vanes. The APU was started to provide hydraulic power (but no air was supplied to the engine). A supervisor suggested that to save time, the person should continue working in the engine as the aircraft was towed to the run bay. The maintainer remained in the engine as the aircraft was towed. There was some disagreement between the crew about whether this was a good idea. About 10 minutes was saved.

Situational Violations

Imagine that you have been asked to check a task recently completed by a trusted colleague. Your signature is needed to confirm that they have performed the task correctly. The work was in an area that can only be seen when various access panels are removed. You arrive ready to inspect their work only to find that your colleague has replaced all the panels. This means that you have no way of knowing what they have done unless you remove the panels. What do you do? Events such as this breed situational violations.

Whereas personal motives—avoiding effort, seeking thrills, showing off, being macho, demonstrating skill and the like—play a large part in shaping routine and thrill-seeking violations, *situational violations* arise from a mismatch between work situations and procedures. The primary aim when committing a situational violation is simply to get the job done. The problem is not so much with the workforce as with the system as a whole. The following railway example will illustrate the point.

The business of a shunter (or brakeman in the US) is to join up railway wagons (see Figure 4.3). In Britain, the rules prohibit shunters from remaining between the wagons during the easing up process—that is, when the shunting engine is pushing the wagons together. The shunters are expected to use a long pole to make the linkage. But sometimes the connecting shackle is too short to make the coupling when the buffers are at full extension. In order to do the job, the shunter has to get between the wagons and hook on the shackle at the moment of contact when the buffers compress. For many, though, this isolated knowledge-based act can become a skill-based or routine way of working. It cuts down on effort and simply being between the wagons seems safe enough, since the buffers are three to four feet apart. Accidents happen when the shunter makes an error while violating—he may slip or become distracted. In the past, there have been a relatively large number of fatalities associated with shunting. In a large proportion of cases, the shunters have died as the result of being crushed between the buffers or falling under the wagon wheels—clearly, these are errors made whilst violating (a violation + error = disaster).[13]

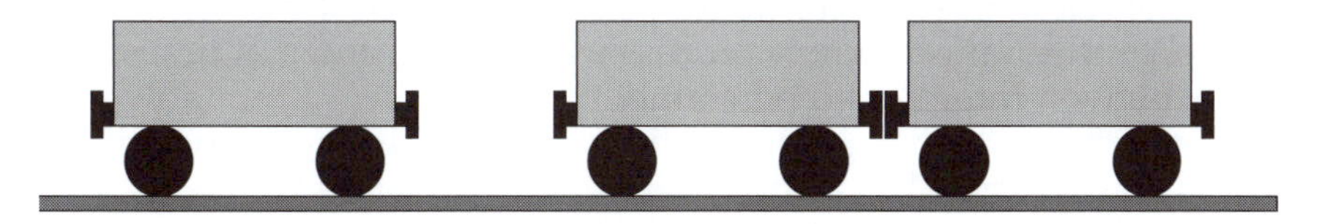

Figure 4.3 The job of the shunter: linking up railway wagons

Maintenance offers many opportunities and temptations for situational violations. Personnel are often faced with the dilemma of being urged by their employers to follow the procedures, while at the same time being encouraged to meet often pressing deadlines. One mechanic summed it up this way: 'Management tell us to follow the procedures to the letter, but then they tell us not to be obstructive and to use common sense'.[14] While many violations may have relatively trivial consequences, some situational violations can be particularly dangerous, especially when they remove defences or safety nets built into the system. In the Australian maintenance survey, over 30 per cent reported that they had decided not to perform a required functional check because of a lack of time.[15]

The following case study illustrates such a scenario.

A Boeing 747 was about to make its first flight after the oil lines on one engine had been changed. After finding oil leaks during the engine run, the maintainers tightened up the connections on the suspect oil lines. They planned to do an additional engine run to check the connections, but the towing tug arrived. The technicians followed the aircraft to the gate where they performed an engine 'dry spin', motoring the engine on the starter. No oil leaks were found. Subsequently, an oil leak from this engine caused an in-flight engine shutdown and the aircraft had to divert. There were no major 'crimes' here, just a few corners shaved off. But these were sufficient to cause an expensive incident.

Situational violations do not only involve people who get their hands dirty. In May 1998, a fire occurred on board the Royal Australian Navy supply ship *Westralia* when a flexible fuel hose burst, spraying diesel fuel over a hot engine.[16] Four of the crew died fighting the fire. The ship had recently undergone maintenance during which rigid fuel pipes had been replaced with flexible hoses. A modification of this type should have been processed through the Navy's formal configuration change process, as well as being approved by Lloyd's Register. However, the formal approval processes were bypassed and unsuitable hoses were fitted. The inquiry found that the Navy's formal configuration change process was circumvented at times, generally by well-intentioned personnel.

The Consequences of Maintenance Errors

Distinguishing between different forms of maintenance error is not just an academic pastime. Understanding the forms of error is important because different errors tend to lead to different consequences. Figure 4.4 is based on the Australian maintenance study, and shows the errors that led to quality incidents (that threatened the safety or operation of an aircraft) alongside the errors that led to worker safety incidents. The most common error leading to quality incidents were memory lapses, accounting for nearly one-third of the errors in this group. Violations and knowledge-based mistakes were the next most frequent errors in quality incidents.

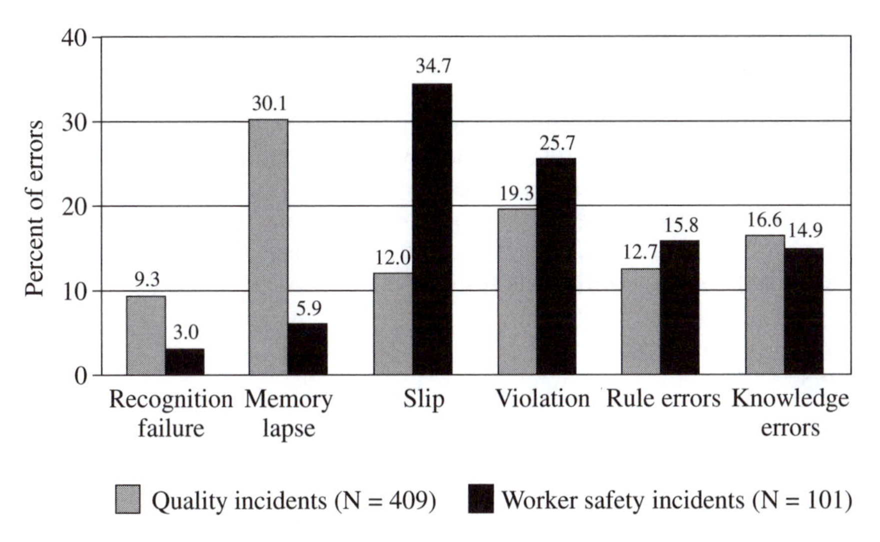

Figure 4.4 Comparison of error types resulting in quality incidents and worker safety incidents

But now compare the picture for worker safety incidents, in which people were injured or otherwise placed in danger while maintaining an aircraft. Here, slips were the most frequent error type, although violations, rule and knowledge-based mistakes were still important. But memory lapses rarely led to health and safety incidents.

The message in a nutshell is that the errors that cause injury to your personnel may be different to the errors that affect the quality of the maintenance work. Both types of outcomes need to be addressed, but different interventions may be required for each. The efforts you are currently putting into health and safety may be making little impact on quality, and vice versa.

Summary

In this chapter, three basic error types were identified—skill-based errors, mistakes and violations. Skill-based errors were divided into three types, recognition failures, memory failures and slips. Mistakes can be divided into two main groups, rule-based and knowledge-based mistakes. Rule-based mistakes typically involve either incorrect assumptions or bad habits. Despite their name, rule-based errors do not generally involve intentional violations of procedures. Knowledge-based errors can reflect failed problem solving or a lack of system knowledge.

Next, we considered the role of violations in maintenance, and we identified three types of violations. These were routine violations, thrill-seeking or optimizing violations and situational violations.

Finally, it must be stressed that while errors and violations can have bad outcomes, they are not intrinsically bad things. Each type relates to normally useful and adaptive mental processes. Nor do they occur in isolation. The next chapter discusses the local workplace factors that set the conditions for errors.

Notes

1 See J. Reason, *Human Error* (New York: Cambridge University Press, 1990) for a more detailed discussion of the nature and varieties of human error.
2 C.G. Drury and J. Watson, 'Human factors good practices in fluorescent penetrant inspection', *Human Factors in Aviation Maintenance* (FAA/Human Factors in Aviation Maintenance, 1999); C.G. Drury and J. Watson, *Human Factors Good Practices in Boroscope Inspection* (FAA/Human Factors in Aviation Maintenance, 2001), <http: / /hfskyway.faa.gov>.
3 Air Accident Investigation Branch, *Report on the Accident to BAC One-Eleven, G-BJRT over Didcot Oxfordshire on 10 June 1990* (London: HMSO, 1992).
4 A. Hobbs, 'The links between errors and error-producing conditions in aircraft maintenance', Paper given to 15th Symposium on Human Factors in Aviation Maintenance, 27–29 March 2001, London.
5 W. James, *The Principles of Psychology, Vol. 1* (New York: Dover Publications, 1890), p. 115.
6 National Transportation Safety Board, *Maintenance Anomaly Resulting in Dragged Engine during Landing Rollout, Northwest Airlines Flight 18, New Tokyo International Airport, March 1 1994*, NTSB/SIR-94/02 (Washington, DC: National Transportation Safety Board, 1995).
7 A. Hidden, *Investigation into the Clapham Junction Railway Accident* (London: HMSO, 1989).

8 A. Hobbs and A. Williamson, 'Skills, rules and knowledge in aircraft maintenance: Errors in context', *Ergonomics* **45** (4), 2002, pp. 290–308.
9 A. Hobbs and A. Williamson, *Aircraft Maintenance Safety Survey – Results* (Canberra: Australian Transport Safety Bureau, 2000).
10 N. McDonald, S. Cromie and C. Daly, 'An organisational approach to human factors', in B.J. Hayward and A.R. Lowe (eds), *Aviation Resource Management* (Aldershot: Ashgate, 2000).
11 Hobbs, op. cit.
12 Ibid.
13 R. Free, 'The Role of Procedural Violations in Railway Accidents', Ph.D Thesis, University of Manchester, 1994.
14. A. Hobbs, 'Maintenance mistakes and system solutions', *Asia Pacific Air Safety*, **21**, 1999, pp. 1–7.
15 Hobbs and Williamson, 2000, op. cit.
16 Department of Defence, *Report of the Board of Inquiry into the Fire in HMAS Westralia on 5 May 1998* (Canberra, Australian Capital Territory, Department of Defence, 1998).

5 Local Error-provoking Factors

In Chapter 2, we saw that human errors do not emerge randomly, but are shaped by situation and task factors that are part of the environment in which the person is functioning. Error-producing conditions in the workplace are commonly referred to as local factors, meaning that they are present in the immediate surroundings at the time of the error. In Figure 5.1, such factors are shown surrounding the error types introduced in Chapter 4. Errors may also reflect wider system problems, deeply rooted in the organization. These 'upstream' issues will be dealt with in later chapters.

There are many potential local factors that can affect worker performance for good or ill. The International Civil Aviation Organization (ICAO) lists over 300 such influences ranging from heat and cold through to boredom, nutritional factors and even dental pain.[1] Ultimately, however, experience shows that a relatively limited

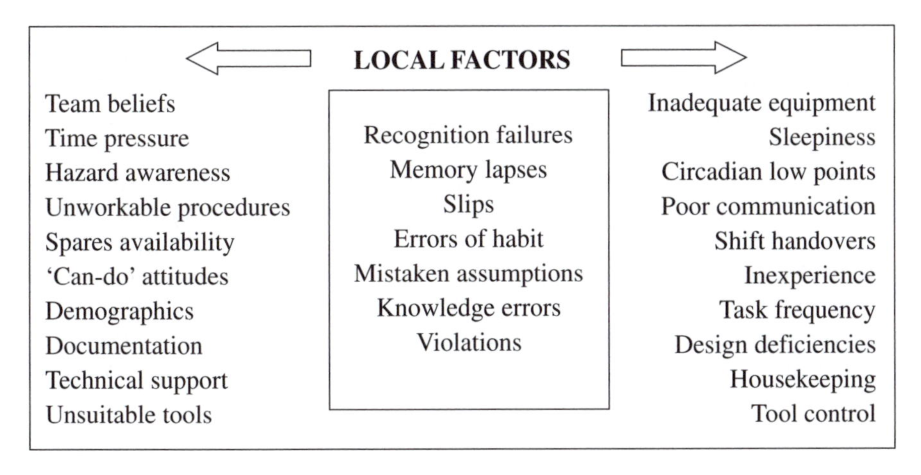

Figure 5.1 Maintenance errors (central box) in the context of local error-producing factors (outer box)

number of local factors appear over and over again in maintenance accident and incident reports.

This means that when considering your own workplace and the factors that increase the probability of error, you can focus on a relatively manageable list of issues. In the following pages, we will consider some of the key factors that surround maintenance work, and the kinds of errors they are likely to promote.

Documentation

Maintenance jobs typically start and finish with documentation. Documents not only convey instructions about task performance, but also play an important part in communication by recording the completion of tasks and the extent of system disturbance.

A study of the normal day-to-day activities of airline maintenance personnel found that for much of the time they were not touching aircraft at all, but were using fiche readers, technical logs, task cards and maintenance manuals, or were signing off tasks.[2] The more unfamiliar the task, the more time was spent dealing with documentation. Documentation may guide performance on new or unfamiliar tasks, but as people become more familiar with a task they are less likely to refer to the paperwork. This has its risks, particularly if procedures change.

Given the importance of paper records in maintenance, it is not surprising that poorly designed documents lie at the heart of many incidents. Procedures that are ambiguous, wordy or repetitive are likely to promote errors. Procedures that are unworkable or unrealistic are likely to promote violations. While re-writing an organization's documentation may not be a feasible short-term objective, some improvements can be made incrementally. Simplified English, for example, can make the language of maintenance documentation clearer and more accessible, particularly in the case of staff for whom English is a second language.[3] Even small improvements in page layout, diagrams and warnings can help to reduce errors. For example, many companies print maintenance documentation in upper case, even though it has been known for many years that such text is more difficult to read than words written in the usual mixture of lower and upper case. Replacing blocks of upper case text with normal mixed-case text can increase reading speed by 14 per cent.[4]

Time Pressure

In aviation, maintenance personnel have faced pressures to get aircraft back into service since the early days of flight. However, as operators strive to reduce the amount of time that aircraft spend out of service, pressure has become a fact of life for most maintainers. A particular risk is that maintenance personnel faced with real or self-imposed time pressures will be tempted to take shortcuts to get an aircraft back into service more quickly.

Maintenance systems have built-in safeguards such as independent inspections and functional tests designed to capture errors on critical tasks. By necessity, these error-capturing safeguards generally occur at the end of jobs, at exactly the time when pressures to get the aircraft back into service are likely to be greatest, and the temptation to leave out or shorten a procedure is strongest.

In the survey of maintenance personnel referred to earlier, time pressure was the most frequently mentioned factor leading to incidents. Of particular concern is that 32 per cent of respondents reported that there had been an occasion when they had not done a required functional check because of a lack of time.[5] At the time, such a decision may have seemed safe and reasonable; however, decisions made under pressure do not always stand the test of hindsight.

Housekeeping and Tool Control

The maintainers who worked on flying boats during the heyday of long-range flying boat services had a simple yet effective way of keeping track of tools. If a task were to be done over water, screwdrivers and other tools would be attached to lengths of twine. Housekeeping, including the way tools and equipment are tracked, is a fundamental local factor that can increase or decrease the chances of errors. It extends to keeping track of items used in maintenance, such as rags and removed or disassembled components.

While military mechanics use tools provided by their unit, stored in controlled shadow-boards or in tool cribs, civilian mechanics the world over usually own the tools of their trade. While stricter tool control may not be feasible as long as workers own their own tools, maintenance can learn from surgery where swab and instrument counts are used to reduce the chances of surgical objects being sewn-up inside the patient.

Ultimately, the housekeeping practices of an organization reflect beliefs about people and how they do their jobs. Leaving removed fasteners in an obvious and prominent location is saying 'I realize that the next person who comes along may not know these have been removed and may need prompting'. Leaving them on a convenient ledge or bench, on the other hand, involves the assumption that people are perfect and will have a complete picture of the situation. Poor housekeeping and tool control practices increase the chances of mistaken assumptions and memory lapses.

The way tools and components are arranged and stored is not just a matter of convenience. It is an important form of communication that provides situational awareness and reduces the chances of error. We will say more about communication in the next section.

Coordination and Communication

While maintenance personnel come with the full range of personal styles, the stereotype of the maintenance person is that of a strong silent type who does the business quietly and without fuss. Some of the most serious maintenance errors have had their origins in poor communication practices.

In a recent survey, senior US maintenance mechanics were asked to identify the most challenging part of their job.[6] Their most frequent answer was 'human relations or dealing with people'. Performing in a team requires more than technical know-how, and we often overlook the need to develop these important communication and people skills. United States NTSB board member John Goglia, who is himself a maintenance technician, has noted that: 'With their engineering focus, maintenance managers and technicians possess highly technical skills, but sometimes lack the communication skills to ensure safety in today's complex operations. What is needed is a better balance of technical skills and social skills.'[7]

In the Australian maintenance survey, 12 per cent of reported occurrences featured coordination problems such as misunderstandings, poor teamwork or communication, or incorrect assumptions.[8] In many cases, coordination breaks down when people make unspoken assumptions about a job, and fail to communicate with one another to confirm the situation. Sometimes, maintainers fear that they will give offence if they are seen to check the work of colleagues too thoroughly or ask too many questions.

The following example illustrates a coordination difficulty that involved unspoken assumptions and poor communication.

> 'Two of us were dispatching the aircraft. The nose steering by-pass pin was left in. Aircraft began to taxi, but stopped as soon as no steering recognised. We removed pin and ops normal. This is a repetitive maintenance task; both of us assumed the other had the pin.'

On other occasions, communication was hampered by 'cultural' barriers and a lack of assertiveness, as illustrated by the following example.

> 'An airframe and engine tradesman did not properly secure an oil line. I did not say anything, though I seem to remember noticing, as electrical instrument maintenance engineer's comments on engine airframe systems are unwelcome/unheeded.'

Most major airlines now provide flight crew with training in non-technical skills such as delegation of tasks, communication, management and leadership.[9] There is an increasing recognition that non-technical skills such as these are as important within maintenance operations as they are for flight crew.[10] More will be said about this in Chapter 8.

Tools and Equipment

Among the most influential local conditions influencing work quality are the tools and equipment available to do the job.

In the Australian survey, the second most commonly cited contributing factor was equipment deficiency, most often a lack of correct ground equipment or tools.[11] For example, a required tool may not have been available, leading to an improvisation. Many of the equipment problems resulted in hazards to maintenance workers themselves.

> 'We had some work to do in the forward cargo compartment. We wanted to get the maintenance done as quickly as possible so an engine stand was used to access the cargo. The top of the stand is about four feet below the floor of the cargo, but was used because it was the only available stand in the area. A person fell out of the compartment onto the stand and then the ground after tripping while exiting the cargo compartment.'

In other cases, the design of ground equipment or tooling had been partly responsible for the incident.

> 'The previous shift had fitted a special tool to the aircraft but had forgotten to remove the tool at the end of the procedure. Also, they had not recorded fitment of the tool in the maintenance log. As I was preparing the aircraft to be towed, I noticed the tool and removed it, but I didn't realise that it was in two halves and, as it was dark, I didn't see that part of the tool was still connected to the aircraft. The problem was noticed when the pilot did his control checks. The tool did not have a streamer attached to it, and was only painted bright red and white after the incident.'

The maintenance of maintenance equipment is itself a crucial task for management, yet one that sometimes does not get the attention it deserves. The very adaptability of maintenance workers is part of the problem. If the right stand is not available, another can be made to fit, if the correct tool is not available, perhaps one can be made. Clearly, equipment deficiencies breed violations, not for the thrill of it, but because there are few alternatives if the job is to be done. If maintainers stopped work when a piece of equipment was not available, the problem would be more obvious to management. But a can-do attitude often prevents this.

Fatigue

Until the industrial revolution, relatively few people worked throughout the night. For thousands of generations, humans have been essentially daytime creatures. As discussed in Chapter 3, a range of bodily functions undergo 24-hour rhythms, linked to the night/day cycle. As night sets in, several changes occur in the human body. Body temperature decreases, the levels of various body chemicals change and most importantly, alertness begins to reduce. Statistics from a range of industries reveal that errors are more likely to occur in the early hours of the morning than at any other time.[12]

Recent research has shown that moderate sleep deprivation of the kind experienced by shift workers can have consequences that are very similar to those produced by alcohol.[13] After 18 hours of being awake, mental and physical performance on many tasks is affected as though the person had a blood alcohol concentration (BAC) of 0.05 per cent. Boring tasks that require a person to detect a rare problem (like some inspection jobs) are most susceptible to fatigue effects. After 23 hours of being continuously awake, people perform as badly on these tasks as people who have a BAC of 0.12 per cent.

One in five of the engineering personnel who responded to the Australian survey said that they had worked a shift of 18 hours or longer in the last year, while some had worked longer than 20 hours at a stretch.[14] There is little doubt that these people's ability to do the job would have been degraded. An important point to note is that like drunks, fatigued individuals are not always aware of the extent to which their capabilities have degraded.[15]

> 'After working approximately 29 hours straight, the last job I had to do was a simple engine component change, one I had done many times before. Following the fitment of the component, I could not focus on the correct rigging procedures. My concentration had lapsed to the point where I could not conduct a simple task.'

Fatigued workers can become more cranky and irritable; but perhaps most importantly for maintenance is the fact that they have trouble controlling their attention. Information slips out of short-term memory more easily, and memory lapses become more likely.

It used to be thought that night workers adjusted and that their body rhythms became inverted, or synchronized so that, for them, the early hours of the morning were like the middle of the day, and the middle of the day was their period of greatest fatigue. We now know, however, that even permanent night-work only results in a general flattening of the 24-hour body cycles. Night workers are not quite as fatigued in the early hours of the morning as a day worker would be, but neither are they able to obtain completely refreshing sleep during the day.

The duration of the shift and the quality of sleep that the person has obtained are also crucial. While some shift-workers claim to be able to get adequate sleep during daytime hours, the sleep obtained during the day is generally briefer and less refreshing than night-time sleep. Maintenance workers may be sleep-deprived at the start of a shift, and the circadian dip in arousal and performance will be even more serious than usual.

Maintenance work at night can present problems other than fatigue. Technical support may be unavailable or else hard to obtain, and supervision may be reduced. There may be no alternative to night-time maintenance, but while fatigue remains a necessary evil, it can be managed. We will discuss fatigue management in Chapter 9.

Knowledge and Experience

A lack of knowledge or experience is one of the most obvious local factors leading to maintenance errors. Most maintenance personnel have had the experience of carrying out a new task, while not being entirely sure whether they were doing it correctly. Such trial-and-error performance is by definition prone to being unreliable. Younger workers, in particular, need to know about the traps lying in wait for them, yet too often they are allowed to discover these for themselves.

The way a maintainer approaches a task will be greatly influenced by whether it is one that he or she has done many times before or is performing for the first time. It is well established, for example, that the time taken to perform a maintenance task decreases the more often it is carried out.[16]

This does not mean that senior maintenance personnel will not make mistakes caused by a lack of experience. In fact, where they have a choice, senior personnel will often seek out the more unusual and challenging tasks. Aviation mechanics spend about 15 per cent of their time on tasks which they have never done before, but senior mechanics spend 20 per cent of their time on such tasks.[17] While routine and boring jobs carry special dangers, including a greater risk of absent-minded slips and lapses, tasks requiring knowledge-based problem solving are much more error prone than tasks that are well understood. This applies regardless of whether it is an apprentice performing a routine task for the first time, or a senior mechanic performing an unusual modification or check.

The lesson for management is that tasks that take workers into unfamiliar territory need to be managed with particular care.

Bad Procedures

Poorly designed procedures are a common source of maintenance error.[18] In the nuclear industry, for example, nearly 70 per cent of all human performance problems have been traced to bad procedures.[19] These were procedures that gave the wrong information, were inappropriate or unworkable in the present situation, were not known about, were out of date, could not be found, could not be understood or simply had not been written to cover this job. Poor procedures not only breed mistakes, they are a major factor leading to violations.

It would be wrong to think that most violations were due to bloody-mindedness on the part of the workforce. As we have already seen, situational or necessary violations arise because people want to get the job done, but the tools or the situation make it impossible to do the job and comply with the procedures. In the study of European airlines, referred to in Chapter 4, it was found that unclear task cards or vague procedures were among the main reasons for deviations from maintenance procedures.[20]

Violations are deliberate acts. People weigh up the costs and benefits of non-compliance and when the perceived benefits exceed the perceived costs, they are likely to violate. This has been called *mental economics*, and the credit and debit sides of the mental 'balance sheet' are shown in Table 5.1.[21]

Table 5.1 The mental 'balance sheet' determining whether or not a person will violate in a particular situation

Perceived benefits	Perceived costs
• Easier way of working	• Accident
• Saves time	• Injury to self or others
• More exciting	• Damage to assets
• Gets the job done	• Costly to repair
• Shows skill	• Sanctions/punishments
• Meets a deadline	• Loss of job/promotion
• Looks macho	• Disapproval of friends

For many acts of non-compliance, experience shows that violating is an easier way of working and brings no obvious bad effects. In short, the benefits of non-compliance are often seen to outweigh the costs. The study of European airline mechanics found that the most common reason for non-compliance with procedures was that there was a more convenient or quicker way of working.[22]

The management challenge here is not so much to increase the costs of violating by stiffer penalties, but to try to increase the perceived benefits of compliance. And that means having procedures that are workable and that describe the quickest and most efficient ways of doing the job. Any lack of trust caused by inappropriate or clumsy procedures will increase the perceived benefits of violating. Indeed, as shown earlier, the job can only be done in some cases by deviating from the procedures, particularly if the formal procedure cannot be followed in the time allowed. Even if everybody knows that the procedures need to be improved, the formal change system may be so slow and unwieldy that it is more expedient to turn a blind eye to the inevitable violations. This has been referred to as the 'double standard of task performance' and is one of the most difficult issues facing maintenance managers.[23]

Procedure Usage

There are many reasons why people choose not to use written procedures—not the least of which is that it is very hard to read and do the job at the same time. It also depends on how the workforce perceives the risks associated with a particular task. Table 5.2 shows the results of a survey of procedure usage in a large petrochemical plant.[24]

Table 5.2 Procedure usage in a petrochemical plant

Task type	Per cent usage
Quality-critical	80 (46)*
Safety-critical	75 (43)
Problem diagnosis	30 (17)
Routine (inc. maintenance)	10 (6)

* Numbers in brackets indicate estimated actual usage given that only 58% of respondents said they had procedures open during task execution.

Source: D. Embrey, 'Creating a procedures culture to minimise risks', Paper given to 12th International Symposium on Human Factors in Aircraft Maintenance, 12–16 May 1998, Gatwick.

The first numbers given show the extent to which the workforce said they used procedures in regard to particular kinds of activity. According to this, safety- and quality-critical jobs were associated with a high usage, while solving problems (even safety-critical ones) and maintenance work involved a much lower usage. The numbers in brackets show estimates of actual usage in these various categories. These estimates are based on the finding that only 58 per cent of the people surveyed (over 4000) said that they have the procedures open and in front them while they are actually carrying out jobs.

In many highly proceduralized industries, it is common for the workforce to write their own procedures as to how jobs should be done. These are jealously guarded and passed on to new members of the work group. They are generally known as 'black books'. The procedure-usage survey found that 56 per cent of operators used informal procedures, as did a startling 51 per cent of managers.

The survey also sought the reasons why people chose not to comply with procedures. The principal factors are listed below.

- If followed to the letter, the job would not get done.
- People are not aware that a procedure exists.
- People prefer to rely on their own skills and experience.
- People assume that they know what is in the procedure.

Personal Beliefs: A Factor Promoting Violations

Unlike mistakes or skill-based errors, violations involve deliberate deviations from procedures or safe practice. Much of what we have learned about why people violate comes from driving research—the roads make an excellent natural laboratory for studying this topic. It is a particularly valuable context for understanding some of the personal factors that lead people into non-compliance.

Research on driving violations suggests that non-compliance is directly related to a number of potentially dangerous beliefs or illusions. Some of the more important of these are outlined below.[25]

- *Illusion of control.* Violators overestimate the extent to which they can govern the outcome of risky situations.
- *Illusion of invulnerability.* Violators underestimate the chances that their rule breaking will lead to bad outcomes. Skill, they believe, will always overcome hazard.
- *Illusion of superiority.* This comes in two forms. First, violators believe themselves to be more skilled than other people. Second, they do not regard their own tendencies to violate as being worse than those of other drivers.
- *I can't help myself.* Violators often feel that the temptation to bend or break the rules is irresistible.
- *There's nothing wrong with it.* Violators do not see their infringements as wrong or dangerous. For example, they rate their speeding behaviour as less annoying, less serious and less risky than do more law-abiding drivers.
- *Everyone does it.* Violators often explain their behaviour by saying that they are simply doing what everyone else does. This is called a *false consensus.* High violators overestimate the proportion of other drivers who also violate.

There is another important belief that relates directly to maintainers.

- *They really expect us to do it.* Maintainers often feel themselves to be in a double bind. They are told not to break the rules, but they are also expected to get the job done quickly. Many resolve this conflict by seeing management's insistence on their compliance as hypocritical: 'They'll turn a blind eye so long as the job gets done quickly, but we can expect little mercy if our violations cause an accident.'

Links Between Errors and Error-provoking Conditions

Having considered the range of error-provoking factors, it is worth bearing in mind that particular local factors tend to bring forth specific types of error. Figure 5.2 illustrates the links between errors and a range of workplace conditions or factors. It is based on an analysis of over 600 maintenance incidents reported in the Australian survey.[26] Errors and factors that are close to each other on the diagram tend to go together. We can see that:

- memory lapses, the most common type of maintenance error, are closely associated with time pressure and fatigue.
- rule-based errors are linked with inadequate procedures and coordination.

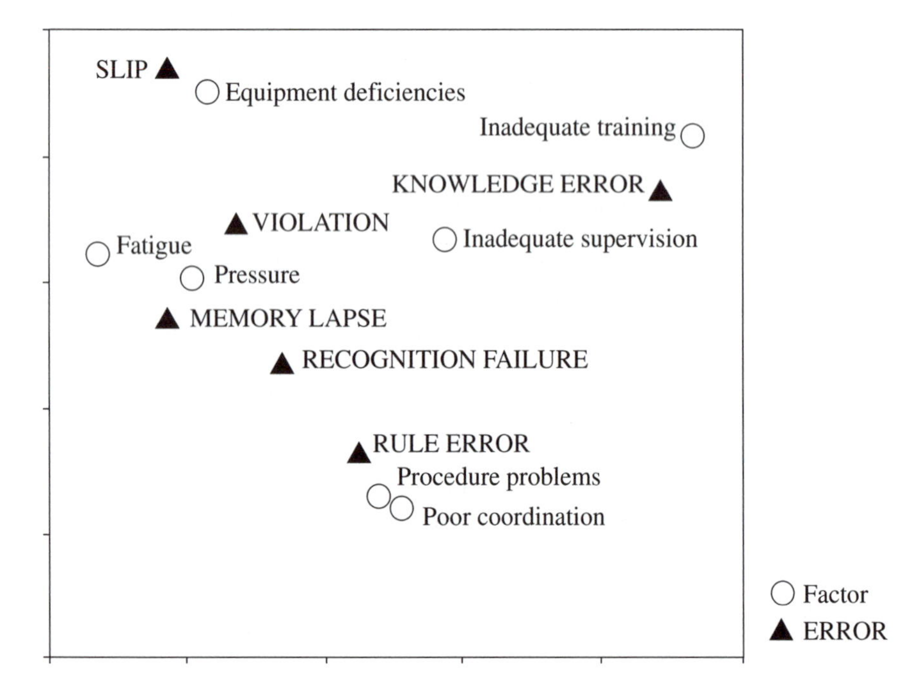

Figure 5.2 Links between errors and contributing factors

- knowledge-based errors show a strong association with training, as we would expect.
- slips are most closely related to equipment deficiencies.
- violations are linked with time pressure. We also know that violations occur in response to inadequate procedures, but this relationship is not shown in this diagram.

Summary

In this chapter, we have introduced the key local factors that are known to increase the frequency of maintenance errors. These include documentation problems, time pressure, poor housekeeping and tool control, inadequate coordination and communication, fatigue, inadequate knowledge and experience and problems with procedures. We then gave particular attention to the role of beliefs in promoting violations. In the next chapter, we examine some case studies of maintenance-related accidents and consider the organizational factors that promote errors.

Notes

1 International Civil Aviation Organization, *Investigation of Human Factors in Accidents and Incidents* (Circular 240-AN/144) (Montreal: International Civil Aviation Organization, 1993).

2 A. Hobbs and A. Williamson, ' Skills, rules and knowledge in aircraft maintenance: Errors in context', *Ergonomics* **45** (4), 2002, pp. 290–308.

3 *AECMA, A guide for the preparation of aircraft maintenance documentation in the international aerospace maintenance language* (Paris: Association Europeenne des Constructeurs de Materiel Aerospatial, 1989).

4 C. Drury, A. Sarac and D. Driscoll, *Documentation Design Aid Development, Human Factors in Aviation Maintenance and Inspection Research Phase VII*, Progress Report (Washington, DC: Federal Aviation Administration, 1997), via <http://hfskyway.faa.gov>. The Document Design Aid provides guidance on the design and documentation of maintenance procedures. It is available at the FAA website.

5 A. Hobbs and A. Williamson, *Aircraft Maintenance Safety Survey – Results* (Canberra: Australian Transport Safety Bureau, 2000).

6 S. Predmore and T. Werner, 'Maintenance human factors and error control', Paper given to 11th Symposium on Human Factors in Aviation Maintenance, 12–13 March 1997, San Diego.

7 J.C. Taylor and T.D. Christensen, *Airline Maintenance Resource Management* (Warrendale, PA: Society of Automotive Engineers, 1998).

8 A. Hobbs, 'The links between errors and error-producing conditions in aircraft maintenance', Paper given to 15th Symposium on Human Factors in Aviation Maintenance, 27–29 March 2001, London.

9 E. Wiener, B. Kanki and R. Helmreich, *Crew Resource Management* (New York: Academic Press, 1993).

10 Taylor and Christensen, op. cit.
11 Hobbs, op. cit.
12 M.M. Mitler, M.A. Carskadon, C.A. Czeisler, D.F. Dinges and R.C. Graeber, 'Catastrophes, sleep and public policy: Consensus report', *Sleep*, **11**, 1988, pp. 100–109.
13 D. Dawson and K. Reid, 'Fatigue and alcohol intoxication have similar effects upon performance', *Nature*, **38**, 17 July 1997, p. 235.
14 Hobbs and Williamson, 2000, op. cit.
15 D.F. Dinges, M.M. Mallis, G. Maislin and J.W. Powell, *Evaluation of Techniques for Ocular Measurement as an Index of Fatigue and as the Basis for Alertness Management* (Washington, DC: National Highway Traffic Safety Administration, 1998).
16 B.S. Dhillon, *Human Reliability with Human Factors* (New York: Pergamon, 1986).
17 Hobbs and Williamson, 2002, op. cit.
18 B. Kanki, D. Walter and V. Dulchinos, 'Operational interventions to maintenance error', Paper given to 9th International Symposium on Aviation Psychology, 27 April–1 May 1997, Columbus, Ohio.
19 INPO, *An Analysis of Root Causes in 1983 and 1984 Significant Event Reports*, INPO 85-027 (Atlanta, GA: Institute of Nuclear Power Operations, 1985).
20 N. McDonald, S. Cromie and C. Daly, 'An organisational approach to human factors', in B.J. Hayward and A.R. Lowe (eds), *Aviation Resource Management* (Aldershot: Ashgate, 2000).
21 W. Battmann and P. Klumb, 'Behavioural economics and compliance with safety regulations', *Safety Science*, **16**, 1993, pp. 35–46.
22 McDonald, Cromie and Daly, op. cit.
23 N. McDonald et al, *Human-Centred Management Guide for Aircraft Maintenance* (Dublin: Trinity College, 2000).
24 D. Embrey, 'Creating a procedures culture to minimise risks', Paper given to 12th International Symposium on Human Factors in Aircraft Maintenance, 12–16 May 1998, Gatwick.
25 J. Reason, D. Parker and R. Free, *Bending the Rules: The Varieties, Origins and Management of Safety Violations* (The Hague: Shell Internationale Petroleum Maatschappij, SIPM-EPO/67, 1993).
26 Hobbs, op. cit.

6 Three System Failures and a Model of Organizational Accidents

Latent Conditions and Active Failures

The technological advances of the last 30 years have made it highly unlikely that a single technical failure or an isolated human error would be enough to cause a major accident. To penetrate a modern industrial system's many defences, barriers and engineered safeguards, now requires the unlikely combination of several contributing factors, each necessary but none sufficient to cause the accident by itself. Exhaustive investigations of accidents in high-technology systems have made it clear that bad events do not usually start at the 'sharp end'. Rather, they involve an interaction between long-standing system weaknesses, termed *latent conditions*, and local triggering events.

Latent conditions are analogous to resident pathogens in the human body. These are disease-causing agencies (viruses, genetic weaknesses and the like) that remain in the system, often for a very long time, doing no obvious harm until they combine with local conditions (various chemicals, diets or life stresses) to breach the body's autoimmune system and so cause illness or death. Most of the ways in which we die these days (for example, cancers and cardiovascular disorders) are due to combinations of these resident pathogens and local triggers.

Latent conditions arise from the strategic decisions made by designers, manufacturers, regulators and top-level managers. These decisions relate to goal setting, scheduling, budgeting, policy, standards, the provision of tools and equipment, and the like. Each of these decisions is likely to have some adverse consequences for some part of the system (under-manning, shortage of resources and so on).

Within the workplace, the local effects of these decisions turn into error- and violation-producing conditions such as those discussed in Chapter 5—time pressure, high workload, the wrong tools, inadequate skills and experience, and so on. These local factors, in turn, interact with human psychology to cause unsafe acts, or *active failures*—errors and violations that have a direct impact upon the system. Such unsafe acts can penetrate some or all of the layers of defence.

It may seem that the latent condition argument is merely an excuse for shifting blame from the maintenance workshop to the boardroom. This is not the case. All high-level decisions, even good ones, will have a downside for someone somewhere within the system. Latent conditions are inevitable in technological systems, just as resident pathogens are always present within the human body.

This chapter looks at three maintenance-involved accidents, each associated with a different technology: an aircraft crash (EMB-120, Texas),[1] a railway collision (Clapham Junction),[2] and an offshore oil and gas platform disaster (*Piper Alpha*).[3] The purpose of these brief case studies is to demonstrate how many different people and events come together to create front-line errors and then allow their damaging effects to go unchecked. Stories like these are a useful way of revealing the far-reaching causal chains that combine to breach a system's defences, barriers and safeguards. The chapter ends by describing a model that links together the various factors involved in the occurrence of an 'organizational accident'.

The Embraer 120 Crash: A Shift Turnover Failure

What Happened

On 11 September 1991, an Embraer 120 (N33701) commercial passenger aircraft suffered a structural break-up in flight and crashed in a cornfield near Eagle Lake, Texas. The three flight crew and 11 passengers were fatally injured. The immediate cause of the disaster was the in-flight loss of a partially secured de-ice boot on the left leading edge of the aircraft's horizontal stabilizer (located on the tail) which led to an immediate and severe nose-down pitch over and the subsequent disintegration of the aircraft. The accident inquiry carried out by the US National Transportation Safety Board (NTSB) revealed that, on the night prior to the accident, the aircraft had received scheduled maintenance which was to have involved the removal and replacement of both the left and right horizontal stabilizer de-ice boots. Investigators at the crash site discovered that the upper attachment screws of the left stabilizer de-ice boot were missing.

In August, the airline had carried out a fleet-wide review of the fitness of aircraft de-ice boots for winter operations. During this review, a quality control inspector had noted that both leading edge de-ice boots on N33701 had dry rotted pinholes along their entire length. They were scheduled for replacement on 10 September, the night before the accident. The work was carried out by two shifts: the second or 'evening' shift and the third or 'midnight' shift. The aircraft was pulled into the hangar during the second shift at around 21.30 hours.

Two second-shift mechanics, with the help of an inspector, used a hydraulic lift work platform to gain access to the T-tail of the aircraft that was about 20 ft above the ground (see Figure 6.1). A second-shift supervisor assigned the job. He then took charge of the work on N33701 (there were two supervisors on the second shift, one dealing with a 'C' check on another aircraft and the one supervising the work on N33701). The two mechanics removed most of the screws on the bottom side of the right leading edge and partially removed the attached de-ice assemblies. Meanwhile, the assisting inspector removed the attaching screws on the top of the right side leading edge and then walked across the T-tail and removed the screws from the top of the left side leading edge.

Figure 6.1 The T-tail assembly on an Embraer 120
The horizontal stabilizer can be seen at the top of the picture

The third-shift hangar supervisor arrived early for work and saw the second-shift inspector lying on the left stabilizer and observed the two mechanics removing the right de-ice boot. He reviewed the second-shift inspector's turnover form and found no write-up on N33701 because the inspector, who had removed the upper screws, had not yet made his log entries. The third-shift supervisor then asked the second-shift supervisor—the one who was dealing with the 'C' check on another aircraft—if work had started on the left stabilizer. The latter looked up at the tail and said 'No.' The third-shift supervisor then told the uninvolved second-shift supervisor that he would complete the work on the right de-ice during his shift, but the work on the left-hand replacement boot would have to wait for another night.

At 22.30 hours, the second-shift inspector (who had removed the upper screws from the leading edges of both stabilizers) filled out the inspector's turnover form with the entry, 'helped the mechanic remove the de-ice boots'. He then clocked out and went home. Later, the inspector stated that he had placed the upper screws removed from leading edges of the stabilizer in a bag and had placed the bag on the hydraulic lift.

The second-shift mechanic (who had removed the right de-ice boot) gave a verbal turnover to the second-shift supervisor (the one working on the unrelated 'C' check). He was told to give his turnover to a third-shift mechanic. This he did and then left. The third-shift mechanic who received the turnover was not subsequently assigned to N33701, though he later recalled seeing a bag of screws on the lift. This mechanic then gave a verbal turnover to yet another mechanic (on the third shift), who later did not remember receiving the turnover; neither did he see the bagged screws.

Yet another third-shift mechanic arrived at the hangar and was told by the third-shift supervisor that he would be working on N33701's right boot replacement. He was instructed to talk to the second-shift supervisor to find out what work had already been done—but he was not told which supervisor to talk to. In the event, he spoke to the one involved in the 'C' check on another aircraft. On asking whether any work had been carried out on the left-hand assembly, the supervisor informed him that he did not think that there would be sufficient time to change the left de-ice boot that night.

The second-shift supervisor actually responsible for N33701 left work at about this time. He did not speak to the other second-shift supervisor, the third-shift hangar supervisor, or to the third-shift supervisor in charge of line checks before he left for home. The second-shift mechanic who had helped with the right boot removal also clocked out and left.

After the shift change, the right leading edge assembly was removed from the horizontal stabilizer by third-shift mechanics. A new de-ice boot was bonded to the front of the leading edge at a workbench. But then N33701 was pushed out of the hangar to make room for work on another aircraft. There was no direct light on N33701 as it stood outside the hangar. After the move, the third-shift mechanics re-installed the right side leading assembly.

A third-shift (quality control) inspector went to the top of the T-tail to help with the installation and to inspect the de-ice lines on the right side of the stabilizer. He told the accident investigators that he did not spot the missing screws on the top left-hand leading edge assembly. He had no reason to expect them to have been removed and the visibility outside the hangar was poor.

Subsequently, the aircraft was cleared for flight. The first flight of the morning passed without incident, except that a passenger later recalled that vibrations had rattled his coffee cup. He asked the flight attendant if he could move to another seat. The passenger did not tell anyone about the vibrations, and the other passengers did not notice them. The accident occurred on the next flight.

Why it Happened

The NTSB's conclusion was that 'the lack of compliance with the General Maintenance Manual by the ... maintenance department led to the return of a non-airworthy airplane to scheduled passenger service'. In short, the shift turnover system was violated. This is an essential system defence when the same job is carried out by more than one shift.

The accident report identified the following 'substandard practices and procedures and oversights' by individual front-line maintainers.

- The second-shift supervisor responsible for N33701 failed to obtain an end-of-shift verbal report from the two mechanics that he assigned to remove both horizontal stabilizer de-ice boots. He also failed to give a turnover to the oncoming third-shift supervisor and to complete a maintenance/inspection shift turnover form. Nor did he give the appropriate work cards to the mechanics so that they could record the work begun, but not completed, by the end of their shift. Had these procedures been carried out, the accident would not have happened.
- The second-shift supervisor not responsible for N33701 (the one in charge of the unrelated 'C' check) informed the third-shift supervisor that no work had been done on the left stabilizer. He had received a verbal turnover from one of the mechanics doing the job, but this came after briefing the third-shift supervisor. On receiving this verbal account of work done, he failed to fill out a maintenance turnover form, nor did he direct his mechanic to brief the other second-shift supervisor responsible for N33701, or to inform the oncoming third-shift supervisor. Instead, he told the mechanic to seek out a third-shift mechanic and report to him what had been done; but this person was not subsequently assigned to the de-ice boot task.

- The second-shift inspector—who had removed the screws on both stabilizers—failed to give a verbal shift turnover to the oncoming third-shift inspector. Moreover, by helping out the two mechanics, he had effectively removed himself from functioning as an inspector.
- The second-shift mechanic, who had assumed responsibility for the work done on N33701 during his shift, failed to give a verbal shift turnover to the second-shift supervisor responsible for the N33701 work. He also failed to obtain and complete the required work cards before leaving at the end of his shift.

Finally, the NTSB considered to what degree senior management should be held responsible for the accident. Three of the board's four participating members concluded that their actions were not causal in this accident. This was disputed by one member, Dr John Lauber, who submitted a dissenting report that included as a probable cause of the accident: 'the failure of (senior) management to establish a corporate culture which encouraged and enforced adherence to approved maintenance and quality assurance procedures.' We will return to this crucial issue of corporate safety culture in Chapter 11.

The Clapham Junction Rail Collision: The Defences that Faded Away

What Happened

At 8.10 a.m. on Monday, 12 December 1988, a crowded commuter train, travelling northwards from Poole to Waterloo in central London, ran into the rear of a stationary train in a cutting just north of Clapham Junction station. After the initial impact, the moving train veered to its right and struck a third oncoming train. Thirty-five people died in the accident and 500 were injured, 69 of them seriously. The immediate cause of the accident was a 'wrong-side' signal failure—one that failed unsafe, or fell on the 'wrong side' of safety. Instead of showing a red aspect, indicating the presence of a train on the next block of line, the signal seen by the driver of the Poole train showed a 'proceed' aspect. This was a new (for the area, at least) four-aspect, non-manual signal whose coloured lights are governed automatically by the movement of trains upon the track.

The work to finish the installation of the new signal, coded WF138, was carried out two weeks earlier on Sunday, 27 November, as part of a massive renovation of the signalling system. The physical preparation for this work—removing an old signal and replacing it with a new one—had been done during the week prior to 27 November, leaving only the connecting of the wires to be completed on the Sunday.

In the previous signalling system, an old wire ran from a relay to a fuse. That relay was labelled DM because it was the track repeater relay (TRR) for circuit DM. Under the instructions for the new system, the circuit was to go from TRR DM to the fuse by a different route, which was to include a further relay TRR DL—the track repeater relay for the track circuit DL. The job scheduled for that Sunday was to connect the new wires and to disconnect the old wire. This disconnection should have been made at both ends of the wire—at the relay end (at TRR RDM) and at the fuse end. In practice, the wire at the fuse end was not disconnected, and although the other (relay end) had been disconnected, it had not been cut back as it should have been, nor was it secured away from its previous contact. Though it had been pushed to one side, it had been left long enough and close enough for it to be possible for it to return to its old position should certain circumstances arise. Tragically, they did.

On the day before the accident (another Sunday), the same technician was again working in the relay room, but with a different assistant. Although not related to the previous wiring work, their job involved changing a relay, TR DN, that—as bad luck would have it—was located immediately to the left of the DM relay on the racking (see Figure 6.2). In the course of working on DN, the position of the old wire on DM was disturbed. This wire, still connected at the fuse end and with its other bare metal end in contact with the terminal, had become a time bomb in the signalling system (see Figure 6.3).

There would have been no disaster on the Monday morning had the trains kept running through the cutting with a sufficient gap between each train. However, the driver of the stationary Basingstoke

Figure 6.2 The locations of track relay DN and track repeater relay DM in the relay room

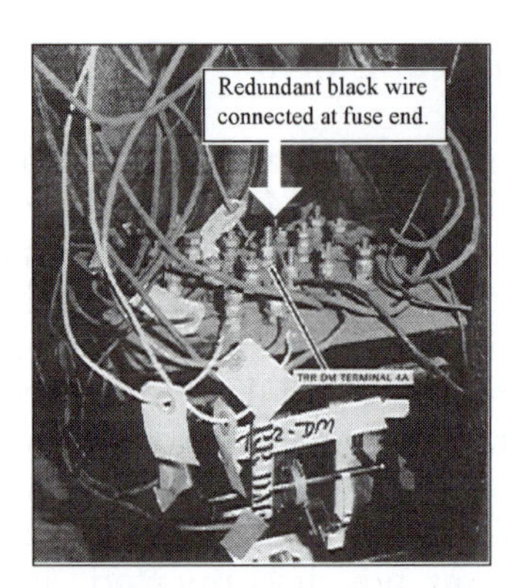

Figure 6.3 Relay TRR DM showing the redundant wire connected at the fuse end
This photograph was taken on the day after the accident

train had stopped to telephone the signalman. The latter believed that WF138 (the one behind the stationary train) had to be at red. But it was not. Because of the rogue wire, it had failed to detect the presence of the stationary train and so gave a 'proceed' aspect to the Poole train.

Why it Happened

The direct causes of the accident—the technician's wiring errors—have already been discussed in Chapter 4. Of more significance here were the system failures that allowed the technician's less-than-ideal working practices to persist uncorrected, and then permitted these highly safety-critical errors to go undetected.

The accident inquiry identified the following contributory system failures.

- Prior to the accident, the technician had taken only one day off in 13 weeks. It was judged that constant weekend working, in addition to the working week, had 'blunted his working edge, his freshness and his concentration'. It was not so much that British Rail (BR) were slave drivers, rather that the technician enjoyed the work and, additionally, welcomed all the overtime he could get. This was a common practice within BR and was well known to management. The technician in question vehemently rejected the suggestion that he was tired on the crucial Sunday—which underlines the point made in Chapter 3 that the subjective and performance effects of fatigue do not necessarily go in step.

- The technician's immediate supervisor failed to monitor and correct the technician's unsatisfactory working habits, as did other supervisors over his 12-year employment with BR. This supervisor also failed to conduct any adequate planning for the weekend work in question, nor did he ensure that the wiring was ready for testing after its completion. He himself was working hard with his gang on the tracks and was not physically able to monitor the wiring technician's activities. He was also judged as having failed to sort out with the testing and commissioning engineer which of them should carry out the testing or how their work should be divided.
- The testing and commissioning engineer, although present in the adjoining signal box on the Sunday in question, made no attempt to carry out an independent wire count (a standard means of quality control), even though he had the necessary time and assistance.
- The supervisor's immediate superior did not choose to work at weekends. He had, however, visited the relay room on an earlier occasion where he saw wires hanging down that were neither cut back nor tied. He did not bring these examples of bad working practice to the attention of either the workforce or management.

And so the story went on up the line of managers. The re-signalling project was running behind time and everyone was extremely busy, many were over-worked. There had also been a major reorganization that year causing a great deal of upheaval. Junior managers were inexperienced in their new jobs, while the roles of the more senior managers changed dramatically. As the Inquiry Report pointed out:

> Poor working practices, unsatisfactory training and incomplete testing had all existed before the reorganisation. The reorganisation did not make any of these factors worse and therefore cannot in that way be seen as part of the cause of the accident. However, the reorganisation could have been an opportunity of coming to grips with the existing system. New brooms might have swept away old bad practices (p. 96).

But no one wanted to rock the boat. And therein lay the root of the problem. High workloads, organizational upheaval, low management morale and a tradition of letting well alone all conspired to allow the system's vital defences simply to fade away. There were no villains in this tragedy (there hardly ever are), just hardworking, over-stretched people trying to do the best job they could under trying circumstances—it sounds like maintenance work anywhere.

The *Piper Alpha* Explosion: Failures of Both the Permit-to-Work and the Shift Handover Systems

What Happened

At 22.00 hours on 6 July 1988, an explosion occurred aboard an oil and gas platform, *Piper Alpha*, located in the North Sea 110 miles northeast of Aberdeen. This led immediately to a large crude oil fire in the oil separation module. The fire extended across into the adjoining module and down to the 68-foot level. The fire was fed initially by a leak from the main oil line to the shore, to which the neighbouring platforms, *Claymore* and *Tartan*, were connected. Very shortly afterwards, a second major explosion—due to a rupture on the riser on the gas pipeline from *Tartan*—caused a terrible intensification of the fire. Flames and smoke engulfed the accommodation area in which the majority of the platform's crew were assembled. There were 226 people on board the platform, of whom 165 lost their lives, together with two crewmembers on a nearby rescue vessel. The

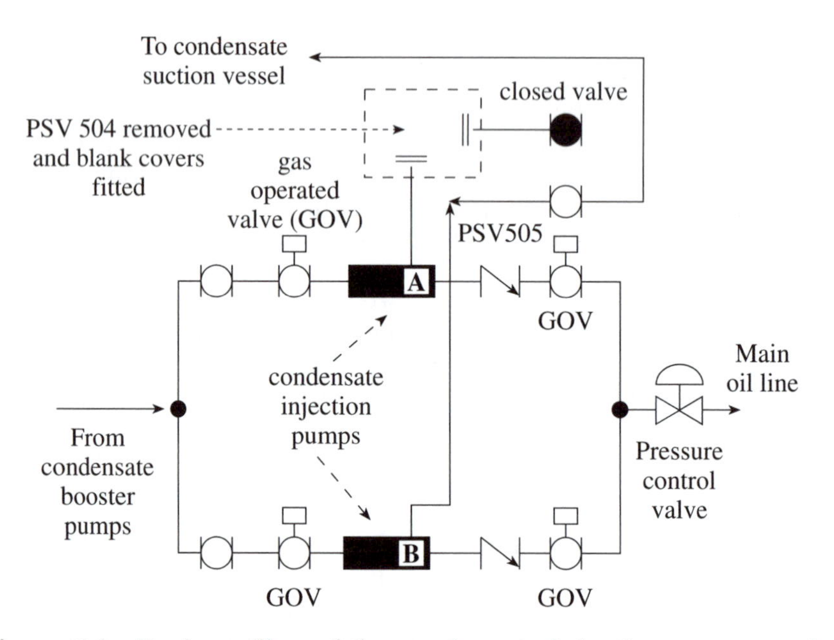

Figure 6.4 Basic outline of the condensate injection pumps and their connections

UK Health and Safety Executive has estimated the total cost of the disaster—less the unquantifiable fatalities—at £2.066 billion. The associated disruption to North Sea production knocked at least a percentage point off the growth rate of output during the ensuing period.

The disaster was triggered at 21.45 hours when one of the two condensate injection pumps (pump B) was tripped by the lead night-shift operator. His intention was to start the other pump (A) that had been shut down for maintenance. The operators were unaware that a pressure safety valve (PSV 504 connected to pump A) had been removed from the relief line of that pump (see Figure 6.4). A blank flange assembly had been fitted at the site of the valve, but it was not leak-proof, allowing condensate to escape. The operators' unawareness of the valve removal resulted from a string of communication failures associated with the permit-to-work and shift handover systems. Many other factors both contributed to and aggravated the disaster, but we will focus here upon the maintenance system failures.

Why it Happened

The two-year long public inquiry, led by Lord Cullen, made the following observations regarding the failure of the maintenance systems.

- The lead night-shift operator would not have attempted to start pump A if he—or the night maintenance lead hand, for that matter—had known that PSV 504 was not in place.
- Information regarding the removal and non-replacement of PSV 504 should have been communicated during the shift handover between the day and night maintenance lead hands. The day maintenance lead hand did not mention this to his night shift opposite number, nor did he record the fact in the maintenance diary or on the A4 pad as the procedures required.
- Pump A was out of service on 6 July. Its pressure safety valve had been removed for recertification. There were some 300 pressure safety valves on *Piper Alpha* and a specialist contractor recertified them every 18 months. The contractor's offshore supervisor, a valve technician (VT), had only been on the platform since 27 June and had received no specific training in the permit-to-work (PTW) system, though he did tell the platform's maintenance superintendent on 28 June that he knew how to work the PTW system. On 6 July, the valve technician came on shift at 06.00 hours and met the day maintenance lead hand who told him that pump A had been shut down and that PSV 504 would be available later that day.

- At 07.00 hours, the VT obtained a PTW form and took it to the maintenance office where the acting superintendent signed it and filled in the tag number, PSV 504, and the location, Module C. From there, the VT went to the Control Room in order to inform the lead operator and get the PTW signed. There was a person sitting at the lead operator's desk who signed the PTW. It later turned out that this person was not the lead operator. The VT and his colleague later removed the PSV and his colleague fitted the blind flanges while the VT began testing the valve. The VT did not check this work.
- At about 18.00 hours, the VT returned to the Control Room to arrange for a crane to lift the valve back down. There was only one person there. The VT did not know who he was, but assumed that he was the oncoming lead operator. (The Inquiry expressed some doubt as to whether this conversation actually took place.) The VT was told that no crane was available and they mutually agreed to suspend the PTW.
- Later, the VT met the day maintenance lead hand and told him that the valve was still off since no crane was available, and confirmed that blind flanges had been fitted.
- Why was the night lead operator unaware that PSV 504 was not in place? There is no clear answer to this question. But even if he had not actually spoken to the VT earlier, the procedures required that he should have signed the suspended PTW and have the job checked by one of his operators. It is clear that this check did not take place. The conclusion of the Inquiry was that neither the night lead operator nor the night maintenance lead hand knew that PSV 504 had been removed because this information had not been transmitted through the handover and PTW systems.
- The Inquiry Report observed that the failure of the PTW system was not an isolated error. The evidence showed that '[The] PTW system was being operated in a casual and unsafe manner. [It] was not being adequately monitored or audited. These were failures for which management were responsible' (p. 231).

The Inquiry Report devoted a chapter to the deficiencies in the PTW and shift handover systems that were judged to have been present prior to the accident. These included the following:

- Lead production operators did not discuss active or suspended permits at the shift changeover. This was required by the company's procedures but had been dropped because there was so much else to do in the time devoted to the changeover briefing.

- Suspended permits were kept in the Safety Office rather than in the Control Room—on the grounds that there was not enough space in the Control Room to display them. A lead operator could be aware of a suspended PTW if it had come to him for signing prior to starting his shift, but he could be unaware of it if it had been suspended some time earlier. There was no general practice for oncoming operators to review suspended permits in the Safety Office before beginning their shifts.
- The training required to make the PTW system work properly was not given, nor was effective monitoring carried out.

The general culture on *Piper Alpha* was one that relied extensively upon informal face-to-face communications between colleagues rather than upon the formal PTW system. There was a general belief among surviving personnel that verbal communications between maintenance and operators were very good. Whether or not this was true, it is clear that on the night of 6 July 1988 the PTW and shift handover systems failed to defend the platform from this appalling tragedy.

As in many disasters, there were prior warnings. On 7 September 1987, a rigger employed by an offshore contractor was killed in Module A of *Piper Alpha* when he fell from an unsecured panel. The company's management admitted that communication failures present within the shift handover and PTW systems had played a significant part in causing this fatality. But such events are only truly a warning if you know what kind of disaster you are going to have. The platform had been visited by an inspector from the Department of Energy's Safety Directorate just 10 days before the disaster. Among other things, his purpose was to check upon the handover and PTW weaknesses brought to light by the rigger's fatal accident. In his report, the inspector stated that the company had 'tidied up' the weaknesses in the PTW and shift handover systems.

Modelling Organizational Accidents

Although the three accidents discussed above occurred in quite different technological systems, they were all organizational accidents. That is, their root causes were present within the system at large. In each case, maintenance errors had been committed, but the origins of these unsafe acts could be traced back to latent conditions within the workplace and in the organization as a whole.

A model describing the contributing factors in an organisational accident is summarized in Figure 6.5. The diagram sets out the basic 'anatomy' of an organizational accident. This model is now used in a variety of hazardous technologies to guide accident investigations

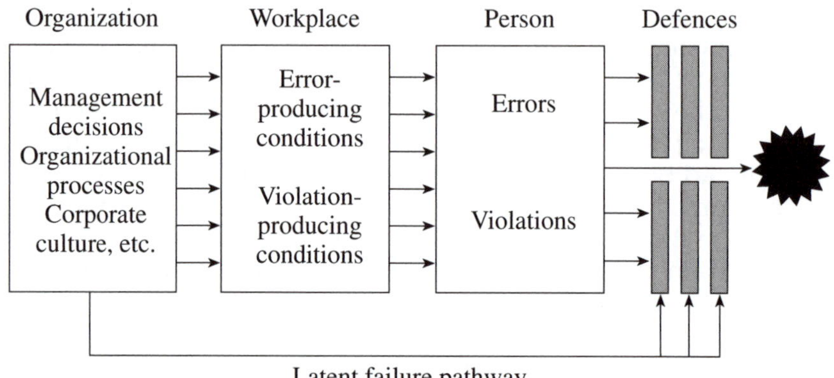

Figure 6.5 A summary of the stages involved in an organizational accident

and to monitor the effectiveness of remedial measures. The direction of causality is from left to right.

- The accident sequence begins with the negative consequences of organizational processes (that is, decisions concerned with planning, forecasting, designing, managing, communicating, budgeting, monitoring, auditing and the like). Another influential factor is the system's culture (see Chapter 11).
- The latent conditions so created are transmitted along departmental and organizational pathways to the various workplaces where they show themselves as conditions that promote errors and violations (for example, high workload, time pressures, inadequate skills and experience, poor equipment and so on).
- At the level of the individual engineer or technician, these local latent conditions combine with psychological error and violation tendencies to create unsafe acts. Many unsafe acts will be committed, but only very few of them will penetrate the many defences and safeguards to produce bad outcomes.
- The fact that engineered safety features, standards, administrative controls, procedures and the like can be deficient due to latent conditions as well as to active failures is shown by the arrow connecting the organizational processes to the defences.

It is clear from the case studies discussed above that the people at the human-system interface—the 'sharp end'—were not so much the instigators of the accident but rather the inheritors of 'accidents-in-waiting'. Their working environments had been unwittingly 'booby-trapped' by system problems.

When systems have many layers of defences, they are largely proof against single failures, either human or technical. The only types of accident they can suffer are organizational accidents; that is, ones involving the unlikely combination of several factors at many different levels of the system that penetrate the various controls, barriers and safeguards.

Defences

Defences, barriers and safeguards are features of the system that have been put there to help the system cope with unplanned and untoward events that have happened in the past, or have been imagined by the system designers. Even the humble pencil comes with an eraser on the end; an acknowledgement that the user is human and is therefore likely to make errors from time to time. An office building provides many examples of defences. Fire is an event that nobody plans to happen; yet we know it is a possible risk. Most buildings have layers of defences against this risk. Some defences are there to provide detection and warning (smoke alarms), some contain the hazards (sprinkler systems), while others allow occupants to escape (fire exits). As well as performing different functions, defences can take a variety of forms. Some are engineered safety features (fireproof doors); others are procedures or instructions (safety signs). The crucial point is that all these features have been designed to cope with, rather than *prevent*, the hazard of fire. Managing the risk of human error is comparable to managing the risk of fire. Nobody wants errors to happen, but we should be prepared for them when they occur. While the responsibility of maintenance managers is to reduce the frequency of maintenance errors, it is also to ensure that the inevitable errors do not take the system beyond tolerable limits.

Some of the most insidious latent conditions are defensive weaknesses. Unfortunately, gaps in defences may only become apparent after an accident has occurred. These gaps may involve defences that have failed to perform as intended or defences that were inadequate or entirely absent.

In regard to human error, maintenance defences serve two important functions: they detect errors and they increase the system's resistance to the consequences of errors.

Defences to Detect Errors

Functional checks and independent inspections are examples of procedural defences designed to uncover previously unknown errors. Unfortunately, procedures are among the weakest forms of defences, as they rely on fallible humans to carry them out. The survey of aircraft maintenance personnel identified that omitted functional checks were one of the most common forms of violations.[4] Independent sign-offs or inspections are also less than perfect methods of detecting error (see also Chapter 12). The reality is that end-of-task inspections are sometimes skipped, particularly if the maintainers consider that the inspection is not necessary, too inconvenient or cannot be performed in the time allowed.

There are other ways in which detection defences can be circumvented, as the following example illustrates. When it became necessary to carry out an unplanned engine change at a foreign airport, one airline wisely decided to send two work crews, one to perform the work, and the other to check into a hotel and rest upon arrival. When the first crew had completed its work, the second crew were then to arrive fresh and rested to perform the necessary functional checks. What happened in practice? The two crews set to work immediately upon arrival and performed the work in half the time. But what was gained in speed was lost by the absence of an independent check.

Defences that Increase the System's Resilience to Error

The second type of defence is designed to contain the consequences of undetected errors. Essentially, this means that we have acknowledged that while we cannot prevent all errors, we can minimize the disturbance they cause. An example is staggered maintenance programming where maintenance of similar or parallel systems is carried out separately to avoid the chance of a single repeated mistake leading to multiple failures of supposedly independent systems.

The special maintenance precautions applied with extended range twin-engine operations (ETOPS) are an example of such an approach.[5] When an aircraft is being maintained in accordance with ETOPS procedures, the performance of identical maintenance actions on multiple elements of critical systems is avoided wherever possible. Engines, fuel systems, fire suppression systems and electrical power are examples of ETOPS critical systems on aircraft such as the B767 and B737. However, these precautions are not generally applied to aircraft with more than two engines, or to twin-engine aircraft that are not being maintained in accordance with an ETOPS maintenance programme.

For example, in 1995, a Boeing 737-400 was forced to divert shortly after departure following a loss of oil quantity and pressure on both engines.[6] Both of the aircraft's CFM-56 engines had been subject to boroscope inspections during the night prior to the incident flight. High-pressure rotor drive covers were not refitted on each engine by mistake and, as a result, nearly all the oil was lost from the engines during the brief flight.

Several months after this event, a similar incident occurred on a Boeing 747-400. Shortly after departing on an over-water flight, the crew noticed reducing oil quantities on the number one and number two engines. The aircraft was turned back to its departure point, where it arrived safely without any need for the engines to be shut down in-flight. After landing, oil could be seen leaking from the engines.

Boroscope inspections had been carried out on all four of the GE CF6 engines. This inspection normally involved removing and then refitting the starter motor from each engine, and the starter motors were removed from the number one and number two engines in preparation for the job. But because the tool to enable the engines to be turned by the starter drive could not be found, the starter motors for engines three and four were not removed and all engines were turned by an alternative method. Due to a lack of spares, a practice had evolved of not replacing O-rings when refitting starter motors. However, on this occasion, a mechanic *did* comply with documented procedures and removed the O-rings from the number one and two starters. However, the workers who refitted the starters apparently assumed that the situation was 'normal' and did not notice that the O-rings were missing—a mistaken assumption.

The incident had a variety of causal factors, such as informal procedures which had evolved to work around the frequent 'nil stock' state of spares, poor lighting and inadequate leak check inspections. However, an important point is that because the aircraft had four engines, it was not protected by ETOPS standards. In essence, the mechanics were 'working without nets'. Had the job proceeded as originally planned, the starter motors would have been removed from all four engines, with potentially serious consequences.

The extension of some ETOPS precautions to non-ETOPS operations would help to contain such maintenance-induced problems. Boeing has encouraged operators as a general practice 'to institute a program by which maintenance on similar or dual systems by the same personnel is avoided on a single maintenance visit'.[7]

Summary

In order to understand how maintenance-related errors come about and to limit their occurrence, we need to go beyond the psychology of the individuals concerned and consider the weaknesses, or latent conditions, existing within the system at large. In short, we need to think about the broader picture.

This chapter examined the maintenance-related contributions to three organizational accidents: the Embraer 120 crash, the Clapham Junction rail collision and the *Piper Alpha* offshore platform explosion. In all three accidents, the principal causal contributions involved deficiencies in the measures designed both to promote accurate situational awareness among the various maintenance personnel working upon the system, and to detect and recover their errors.

A model of accident causation was outlined that involved a cascade of contributing influences from organizational factors, to the resultant error-provoking conditions in the workplace, to the commission of errors and violations at the 'sharp end' and finally to the breached defences that allowed hazards to come into damaging contact with victims. Two kinds of defences in maintenance were then considered: those designed to detect error and those intended to increase the system's error tolerance.

In the next chapter, we will take the lessons learned from these and other tragedies to formulate a set of principles for achieving effective error management in maintenance-related activities. It is now clear that these must include measures that are directed at reducing the error-provoking properties of tasks, workplaces and organizations, as well as those designed to limit individual errors and violations. Most active failures—errors and violations at the 'sharp end'—tend to have short-lived consequences; latent conditions, on the other hand, if left undetected and uncorrected, can continue to create future errors, as well as leaving long-term gaps in the system's defences.

Notes

1 National Transportation Safety Board, *Britt Airways, Inc. Continental Express Flight 2574 In-Flight Structural Breakup EMB-120RT, N33701, Eagle Lake, Texas, September 11, 1991* (Washington, DC: NTSB, 1991).

2 A. Hidden, *Investigation into the Clapham Junction Railway Accident* (London: HMSO, 1989).

3 The Hon. Lord Cullen, *Public Inquiry into the Piper Alpha Disaster* (London: HMSO, 1990).

4 A. Hobbs and A. Williamson, *Aircraft Maintenance Safety Survey – Results* (Canberra: Australian Transport Safety Bureau, 2000).

5 Federal Aviation Administration, Advisory Circular 120 42A, *Extended Range Operation with Two-Engine Airplanes (ETOPS)* (Washington, DC: FAA, 1988).

6 Air Accident Investigation Branch, *Report on Incident to Boeing 737-400, G-OBMM, Near Daventry on 23 February 1995*, Department of Transport (London: HMSO, 1996).

7 Boeing Service Letters, Dual System Maintenance Recommendations, 17 July 1995. See also B.J. Crotty, 'Simultaneous engine maintenance increases operating risks, *Aviation Mechanics Bulletin*, **47** (5), September–October 1999, pp. 1–7.

7 Principles of Error Management

Nothing New

There is nothing new about trying to manage error. All responsible organizations involved in hazardous operations have long employed a wide variety of error management (EM) measures. In maintenance organizations, these include:

- selection
- training and retraining
- work planning
- job cards
- tags and reminders
- shift handover procedures
- licence-to-work systems
- human resource management
- licensing and certification
- checking and sign-offs
- technical and quality audits
- procedures, manuals, rules and regulations
- disciplinary procedures
- Total Quality Management (TQM).

These techniques have evolved over many decades. Though some are tried and tested, they have collectively failed to prevent a steady rise in maintenance-related errors.[1] Their limitations include being piecemeal rather than principled, reactive rather than proactive, and fashion-driven rather than theory-driven. They also ignore the substantial developments that have occurred over the last 20 years in understanding the nature and varieties of human error.

This brief chapter sets out the fundamental principles of error management. Some reiterate what has already been stated in previous chapters. Others anticipate what will be discussed in later chapters. The aim is to provide a concise summary of the error management philosophy presented in this book. Some readers with a very practical turn of mind may shy away from the term 'philosophy'. But all management depends essentially upon what Earl Wiener has termed the 'Four P's': philosophy, policy, procedures and practices.[2] In the case of error management, philosophy is worth at least double the value of each of the other three. Without a unifying set of guiding principles, our efforts will have but a small chance of success.

The Principles of Error Management

Human Error is Both Universal and Inevitable

Human error is not a moral issue. The consequences of errors may be undesirable, even destructive, but making them is as much a part of human life as breathing and sleeping. We include here violations, which are rarely malicious acts and are often well intentioned. Human fallibility can be moderated, but it can never be eliminated. Nor should it be.

Errors are not Intrinsically Bad

Success and failure spring from the same psychological roots. Errors are fundamentally useful and adaptive things. We are error-guided creatures. Errors mark the boundaries of the path towards successful action just as landing lights define the edges of a runway. Without them, we could neither learn nor acquire the skills that are essential to safe and efficient work.

You cannot Change the Human Condition, but You can Change the Conditions in which Humans Work

The problem with errors is not the psychological processes that shape them, but the man-made and sometimes unforgiving workplaces that exist within complex systems. There are two parts to an error: a mental state and a situation. States of mind like moments of inattention or occasional forgetfulness are givens, but situations are not. And situations vary enormously in their capacity for provoking unwanted actions. Identifying these error traps and recognizing their characteristics are essential preliminaries to effective error management.

The Best People can make the Worst Mistakes

A common belief is that a few inept individuals are responsible for most of the errors. If this were the case, the solution to the error problem would be relatively simple: identify these people and then retrain them, sack them or promote them out of harm's way. But the record suggests that the reality is quite different. Some of the world's worst maintenance-related accidents have been due to errors committed by highly experienced people with 30-year blameless records. Errors can strike anywhere at any time. No one is immune. And it is also the case that the best people often occupy the most responsible positions so that their errors can have the greatest impact upon the system at large.

People cannot Easily Avoid those Actions They did not Intend to Commit

Blaming people for their errors is emotionally satisfying but remedially useless. Moral judgements are only appropriate when the actions go as intended and the intention is reprehensible. Blame and punishment make no sense at all when the intention is a good one, but the actions do not go as planned. We should not, however, confuse the issues of blame and accountability. Everyone ought to be accountable for his or her errors. If the error maker does not acknowledge the error and strive to avoid its recurrence, then no lesson has been learned and little or nothing gained from the experience.

Errors are Consequences rather than Causes

The natural human tendency after a bad event is to track back in time to the first deviant human action and call it the cause.[3] We then go on to say that Person X caused the event and punish him or her accordingly—often in proportion to the extent of the damage or injury. This may be appropriate in societies that operate by the 'eye for an eye' principle, but it is totally out of place in maintenance organizations in which accidents arise from the complex interaction of many different factors and where the primary aim of any subsequent inquiry should be to strengthen the system's defences. From this perspective, errors are best regarded as consequences rather than causes. Just like the bad event, errors have a history. Each is the product of a chain of events that involves people, teams, tasks, workplaces and organizational factors. Discovering an error is the beginning of the search for causes, not the end. Only by understanding the circumstances that gave rise to them can we hope to limit the chances of their recurrence.

Many Errors Fall into Recurrent Patterns

Errors can arise from either a unique combination of circumstances, or from work situations that recur many times in the course of maintenance-related activities. The former are random errors—in the sense that their occurrence is very hard to foresee—while the latter are systematic or recurrent errors. As we saw in Chapter 1, more than half of the human factors incidents in maintenance are recognized as having occurred before, often many times.[4] We also observed that certain aspects of maintenance, particularly reassembly or reinstallation, commonly give rise to particular kinds of errors, notably omissions—leaving out essential steps in a sequence or failing to remove unwanted objects on completion. Another frequent group of errors involves miscommunication or lack of communication both within and between maintenance teams or shifts. Targeting these recurrent error types is the most effective way of deploying limited EM resources.

Safety-significant Errors can occur at all Levels of the System

Making errors is not the monopoly of those who get their hands dirty. Managers often think that error reduction and error containment is something that applies exclusively to the workforce, the people at the 'sharp end'. A general rule of thumb, however, is that the higher up the organization an individual is, the more dangerous are his or her errors. Error management techniques need to be applied across the whole system.

Error Management is about Managing the Manageable

One of the commonest errors in error management is striving to control the uncontrollable. Most obviously, this entails trying to change those aspects of human nature that are virtually unchangeable—that is, our proneness to distraction, preoccupation, moments of inattention and occasional forgetting. And, when these attempts fail (as they surely will), the next mistake is to try to shift blame and responsibility away from the company at large and on to those unfortunates known to have made errors. It is the prevalence of these misguided approaches that has led, in large part, to the steady increase in public awareness of maintenance-related errors.

Situations, and even systems, are manageable; human nature—in its broadest sense—is not. Most of the enduring solutions to human factors problems involve technical, procedural and organizational measures rather than purely psychological ones. The problem is that blaming individuals rather than reforming systems has its roots deep in human nature. Such a response may offer some short-term emotional satisfaction, but it provides little or nothing in the way of lasting improvements. An important step in effective error management is to recognize the existence of this *fundamental attribution error* (as the psychologists call this person-blaming tendency) and to fight against it.[5]

Error Management is about Making Good People Excellent

The assumption is often made that error management is a process for making a few error-prone people better. This is not the case. The principal aim of EM to is to make well trained and highly motivated people excellent. Excellence in any professional activity has two crucial elements: technical skills and mental skills. Both need to be acquired through training and practice. Several studies have shown that the mental element is at least as important as the possession of the requisite technical skills. The mental element involves a number of different components. Among the most important of these is mental readiness. Excellent performers routinely prepare themselves for potentially challenging activities by mentally rehearsing their responses to a variety of imagined situations. To do this effectively requires an understanding of the ways in which a task can go wrong. In the case of maintenance-related tasks, this means having an awareness of the varieties of human error and the situations that provoke them. Self-knowledge of this kind is at least as important as anticipating the ways in which a piece of equipment can fail.

Maintainers need to be trained in two ways. First, they need to be informed about the ways in which human performance problems can arise and to be aware of their recurrent patterns. In other words, they need to be familiar with the material discussed in Chapters 3 and 4. Second, they must acquire the skill of mental preparedness. That is, they should be trained to approach each task by mentally rehearsing the ways in which they and their colleagues could go wrong. This not only alerts them to risks of error-provoking situations, but also allows them to plan how they might detect and recover these errors before they cause harm. Improving the skills of error detection is at least as important as making people aware of how errors arise in the first place.

There is no One Best Way

This principle applies in two ways. First, there is no one best technique for error management. Different types of human factors problems occur at different levels of the organization and require different management techniques. Mistakes and violations, for example, have different underlying mechanisms. Mistakes are mainly information-processing problems, while violations have a large social and motivational component. Effective or *comprehensive* error management involves targeting different counter-measures at different parts of the company: the person, the team, the task, the workplace and the organization as a whole.

Second, there is no one best package of EM measures. Different organizational cultures require the 'mixing and matching' of different combinations of techniques. Methods that work well in one company can fail in another, and conversely. This is why these EM principles are so important. There are many ways of achieving the principled reduction and containment of human factors problems. It is up to each organization to choose or to develop the methods that work best for them. The next chapters provide some idea as to the types of error management techniques already available. But it is often better to devise your own methods, or at least to adapt existing techniques to suit your local needs. People are often more likely to 'buy in' to homegrown measures than to ones that have been imported from elsewhere. So long as the basic principles are understood and followed, there are many different ways of achieving effective EM—and most of them have yet to be devised.

Effective EM aims at Continuous Reform rather than Local Fixes

There is always a strong temptation to focus upon the last few incidents and to try to make sure that they, at least, will not happen again. This tendency is further strengthened by the engineer's natural inclination to solve specific concrete problems. But trying to prevent the recurrence of individual errors is like swatting mosquitoes. You kill one and the rest keep coming to bite you. As pointed out in Chapter 2, the only way to solve the mosquito problem is to drain the swamps in which they breed. In the case of maintenance errors, this means reforming the conditions under which people work as well as strengthening and extending the system's defences. Reform of the system as a whole must be a continuous process whose aim is to reduce and contain whole groups of errors rather than single blunders.

The Management of Error Management

EM has three components: error reduction, error containment and managing these activities so that they continue to work effectively. Of these three, the latter is by far the most challenging and difficult task. For EM to have a lasting effect, it needs to be continuously monitored and adjusted to changing conditions. It is simply not possible to order in a package of EM measures, implement them and then expect them to work without any further attention. You cannot just put them in place and then tick them off as another job completed. Here, the bulk of the effort lies in the process rather than the product. In an important sense, the process—the continuous striving toward system reform—*is* the product. We will consider these management issues in further detail in Chapter 12.

Summarizing the EM Principles

1. Human error is both universal and inevitable.
2. Errors are not intrinsically bad.
3. You cannot change the human condition, but you can change the conditions in which humans work.
4. The best people can make the worst mistakes.
5. People cannot easily avoid those actions they did not intend to commit.
6. Errors are consequences rather than causes.
7. Many errors fall into recurrent patterns.
8. Safety-significant errors can occur at all levels of the system.
9. Error management is about managing the manageable.
10. Error management is about making good people excellent.
11. There is no one best way.
12. Effective error management aims at continuous reform rather than local fixes.
13. Managing error management is the most challenging and difficult part of the EM process.

Notes

1. An ICAO Human Factors Digest estimated that the annual average number of maintenance-related air accidents had increased by more that 100 per cent over the preceding decade, while the average number of flights had increased by less than 55 per cent. See D. Maurino, *Human Factors in Aircraft Maintenance*, ICAO Human Factors Digest No.12 (Montreal: International Civil Aviation Organization, 1994).
2. A. Degani and E. Wiener, 'Philosophy, policies, procedures and practice: The four "P's" of flight deck operations', in N. Johnston, N. MacDonald and R. Fuller (eds), *Aviation Psychology in Practice* (Aldershot: Avebury, 1994).
3. H.L.A. Hart and A. Honoré, *Causation in the Law* (2nd edn) (Oxford: Clarendon Press, 1985).
4. A. Hobbs, *Human Factors in Aircraft Maintenance: A Study of Incidence Reports* (Canberra: Bureau of Air Safety Investigation, 1997).
5. S.T. Fiske and S.E. Taylor, *Social Cognition* (Reading, MA: Addison-Wesley, 1984).

8 Person and Team Measures

As technology advances and systems become more complex, we sometimes forget that even the most sophisticated processes still rely on human judgement and skills, whether in medicine, transport, manufacturing or other industries. We can redesign tools, select better materials and develop better processes. But we cannot design better people. Whether we like it or not, we are stuck with the original 'Mark I' human being to maintain our complex systems. And an unchangeable part of the human condition is fallibility.

This chapter is about error management strategies directed at the individual maintainer and at the work team. Although, as indicated in Chapter 6, many maintenance incidents reflect system problems, the maintainers themselves are the last line of defence in that system. Most workers have little control over system issues such as hours of work, equipment and work schedules, but there are things that individuals can do to reduce their chances of being involved in an incident. Most importantly, our skills, habits, beliefs and knowledge can all be changed in ways that increase the reliability of human performance.

Person Measures

Understanding Error-provoking Factors

The first step for maintenance personnel is to obtain some basic knowledge about human performance along the lines of the material in Chapters 3 and 4. Maintainers need to know about the limitations of their short-term memory, how fatigue affects performance, and a host of other facts about human strengths and weaknesses. For this reason, human factors training for maintenance personnel is now required by the International Civil Aviation Organization (ICAO) and the European Joint Aviation Authorities (JAA).[1]

Once maintainers are aware of their own vulnerabilities, they can learn to recognize human performance danger signs.[2] The ability to appreciate the significance of these 'red flags' is a self-protecting skill worth developing. Below are listed some of the most potent error-provoking factors in maintenance activities.

Excessive reliance on memory

Our memories are not always as reliable as we think, particularly when we are tired. Memory lapses are the most common errors in maintenance. It is tempting fate to interrupt a part-completed job without adequate reminders to tell you, and others, of its stage of progress.

You run the risk of a memory lapse every time you try to keep a critical task step in mind to perform later without any reminders. It is better to assume that you will forget, and take precautions, than to hope that you will remember.

Interruptions

Maintenance activities are subject to frequent interruptions—someone needs advice, there is a phone call, or an urgent job needs to be finished off elsewhere. Whatever their nature, however, all such interruptions act to raise your stress level and increase the likelihood of a memory lapse. The most likely error is an omission. It is very important that you are aware of these risks and take steps to combat them. An obvious counter-measure is to anticipate the 'now where was I?' question when you take up the task again, and to leave behind a clear reminder of exactly where you had to stop. We will say more about omission management and reminders in the next chapter.

Pressure

Signs of pressure come in many forms. Being asked repeatedly 'How long is it going to take?' Getting angry during a job. Starting to curse more than usual. Being anxious to go home as soon as you complete the job. Under such pressures, even the most careful workers can find themselves leaving out steps or taking shortcuts. Be aware of these pressures and ensure that they do not lead to risk-taking or cutting corners.

Tiredness

You may not feel tired, but if you have not had a good night-time sleep in the last 24 hours, or you have been at work for longer than 12 hours there is a good chance that you will be impaired by fatigue. Fatigue can increase your chances of making errors, particularly memory lapses. Sleepy people are also more irritable and harder to work with.

Inadequate coordination between maintainers

A breakdown in coordination is one of the most common circumstances leading to incidents. In many cases, coordination breaks down when people make unspoken assumptions about a job, without actually communicating with each other to confirm the situation. Sometimes maintainers fear that they will give offence if they are seen to check the work of colleagues too thoroughly or ask too many questions.

Coordination danger signs include rushed shift handovers, a lack of adequate communication, not asking questions because you feel silly or do not want to offend a work colleague, and working with unfamiliar people.

Unfamiliar jobs

If you are performing a task that is not part of your normal duties, even if you used to perform it in previous years, you are entering a danger zone for error. If an unfamiliar maintenance task is being performed on the basis of 'trial and error', you must recognize that your chances of getting it wrong are greatly increased. A further point to note is that a significant number of aviation maintenance incidents have involved supervisors helping out by getting involved in hands-on work. Although such people may be technically qualified to perform the work, and indeed highly motivated to do a proper job, their practical skills may be degraded.

Ambiguity

Any situation in which you are unsure of what is going on should be a sign to call a halt and clarify the task. Such situations are particularly common in team-based work environments, where 'diffusion of responsibility' can result in people assuming that someone else knows what is going on and has taken charge.

Highly routine procedures

Any procedure that you can perform 'with your eyes closed', such as opening and closing access covers or checking oil levels, is a danger zone for slips and lapses. Because we are so familiar with such tasks, our attention may wander elsewhere, leaving our actions largely under the control of the 'mental autopilot'. While we cannot stop such tasks from being 'on automatic', we can remain vigilant and spot the errors that will occur from time to time.

Understanding Why People Violate Good Procedures

Unlike mistakes or lapses, violations are—for the most part—deliberate actions. People *choose* to deviate from safe operating procedures, rules and accepted codes of practice. This intentional element makes them far more amenable to management at the individual and team levels. Changing the underlying attitudes and beliefs is often a slow business, but it can be done.

Intentions to violate are shaped by three interrelated factors:[3]

- *Attitudes to behaviour ('I can do it.').* These are the beliefs a person has regarding the consequences of some behaviour. How do the perceived advantages of violating balance out against the possible risks and penalties?
- *Subjective norms ('They would/would not like me doing this.').* These are the views that some important reference group (relatives, colleagues, friends and so on) might hold about your behaviour. Will they approve or disapprove, and how much does the person want to be admired or respected by these 'close others'?
- *Perceived behavioural control ('I can't help it.').* How much control does the person feel that he or she exercises over the violating behaviour? This factor is likely to be of considerable importance in fatalistic cultures, particularly in regard to judgements about the consequences of violations. It will also come into play when the individual observes that although local management may pay lip service to the regulations, they actually condone violations, particularly where non-compliance means getting an urgent job done on time. Where the local climate neither rewards compliance nor sanctions violations, the individual is likely to feel that he or she has little control over the situation and can best fit in by doing what everyone else does.

Measures Designed to Reduce Violations

The most commonly used technique for reducing violations is aimed at the first of the three elements above, *attitudes to behaviour*. Efforts are made to scare people into compliance by graphic posters and videos that highlight the grisly consequences of unsafe behaviour.

These fear appeals can serve three useful functions. First, they inform people about the links between unsafe acts and safety. Second, they have the power to shock. Third, they should indicate the correct way of doing things.

However, fear appeals, by themselves, tend to have only limited impact. The main problem is that those most likely to violate—young men—tend to be immunized against such appeals by their unrealistic optimism.

Social controls address the second factor listed above: *subjective norms*. That is, the extent to which other people—whose opinions matter to the individual—would approve or disapprove of the violating behaviour.

The evidence to date is that social controls represent one of the most effective ways of modifying individual behaviour. We are social animals. We need the approval, liking and respect of those we care about. If we become convinced that these 'significant others' will strongly disapprove of our planned actions, we are likely to think twice before carrying out this behaviour. This does not mean that we will never violate, but it is likely to give us pause.

Many of the manageable ways of influencing social controls involve group discussions and group activities. One technique that has been used with considerable success to modify the driving behaviour of Swedish company drivers entails a three-stage procedure, summarized below.[4]

Stage 1 Individuals are sorted into small groups. Ideally, this should be done on the basis of shared work and locations. Each group has a trained discussion leader or facilitator. The first session is taken up with discussing the general quality and safety problems that these people encounter during the course of their work. The discussion leaders then list the items raised.

Stage 2 At their second meeting, the same groups go through the list of problems raised in the first session. The main task is to divide the problems into two groups: those that must be passed up to management to be solved, and those that the group members feel that they themselves could usefully tackle. Violating behaviour and other unsafe acts are clearly items that the group members should be prepared to 'own'. Other items are passed on to management.

Stage 3 At the final meeting, each group focuses on those problems that the members felt they could tackle. They collectively discuss how each problem could be dealt with. Then, at the end of the session, each group member writes down what he or she is going to do about solving these problems (that is, what they themselves could practise as from that moment). They are not required to show this resolve to anyone else. It is simply a personal reminder of their resolutions.

The Swedish study using this technique had two very interesting findings. First, the 850 drivers involved in the discussion groups showed a 50 per cent reduction in driving accidents—accidents that, for the most part, are caused by errors made while violating. The control group (an equivalent number of drivers matched on all fac-

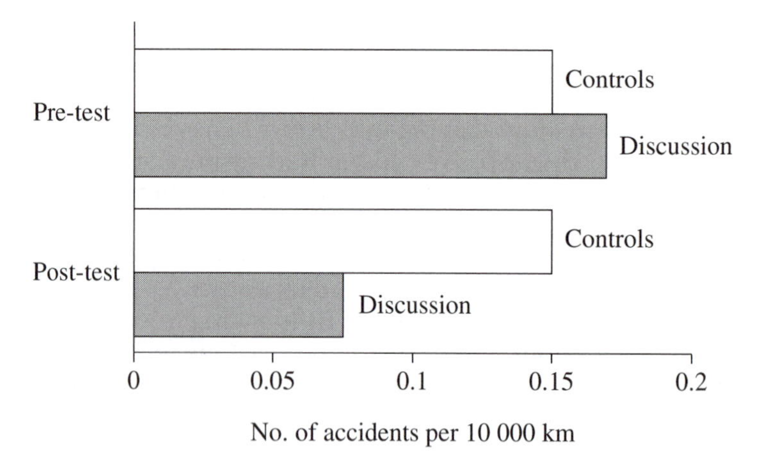

Figure 8.1 Comparing accidents per 10 000 km for control and discussion groups pre-test and post-test

Source: B. Brehmer, N.-P. Gregersen and B. Moren, *Group Methods in Safety Work* (Uppsala: Institute of Psychology, Uppsala University, 1991).

tors except that they did not participate in discussion groups) showed no change over the five-year period of the study (three pre-test years and two post-test years). The results are summarized in Figure 8.1.

The second interesting finding was that most of the drivers involved in the discussion groups did not believe that these activities had played any significant part in this reduction (though the results outlined above clearly showed that they had). A questionnaire sent out after the discussions revealed that only 12 per cent of the respondents considered that their driving styles had changed as a result of their participation, and only 25 per cent thought that the group exercises had any beneficial effects.

How can we explain this? The Swedish drivers already possessed a positive attitude to safety, though—since safety was rarely a topic of conversation between them—they were not aware of how other drivers felt. The value of the group discussions lay in making these group values visible to all members of the discussion groups. The second effect of the discussions was to link the motivation to drive safely with actual decisions about how this could be achieved. Prior intentions to perform a task safely and reliably have a powerful impact on the subsequent behaviour. It is also likely that this impact is considerably amplified by intentions that are formed in a group setting—even though these intentions may not actually be shared with other members of the group.

There are two important conclusions from this and other studies showing similar beneficial effects. First, group discussions of this kind can be very effective in creating safer and more reliable behaviour. They seem especially well-suited to the maintenance environment, and it is not difficult to see how their basic structure could be adapted to reduce violations.

Second, people do not always understand why their behaviour has changed. There is additional research evidence to show that people will not report the real reasons for a behaviour change when this conflicts with their ideas of what constitutes an effective cause. Instead, they report their ideas or theories of why things changed. The moral is clear: do not always believe what people say about why they have changed their behaviour.

It must be stressed that changing attitudes is a slow and difficult business. The goal in any such programme is to reach a point where the control of compliant behaviour shifts from external to internal factors, from externally imposed rewards and sanctions to intrinsic motivation, where the individual simply prefers to comply.

The route from external to internal control passes through two intermediate stages. The first of these is *guilt-driven*: the person obeys the dictates of conscience because he or she seeks to avoid guilt (as opposed to wanting to follow the rules). The next phase is *identification*. Here, the person identifies with the outcome of the compliant behaviour. He or she may not like what this entails, but enjoys the 'feel good' factor when compliant performance has been achieved (that is, we may not like the idea of answering a heap of overdue mail, but are pleased with the results). Only at the final point of *intrinsic motivation* is there no internal conflict.[5]

Studies of addiction suggest that there are five stages in breaking a bad habit. The first stage is *pre-contemplative*, the person has not even considered breaking the habit. The second is *contemplative*, the person begins to feel that it might be a good idea to become free of the habit but has made no firm decision to change. The third stage is *making the decision*; here the person truly resolves to give up the habit. The fourth stage is *actually stopping*, and the final stage has to do with procedures designed to *maintain* the improved state.

Notice that actual behavioural changes do not occur until quite late on in the process. Many techniques seeking to change behaviour might appear to have disappointing results in objective terms, but they may have achieved significant subjective progress in prompting the crucial decision to change (see Figure 8.2). Do not underestimate the importance of moving through these preliminary stages before you can expect to see real results. Remember also that people are not always able or willing to tell you where they currently stand along this scale. In short: *keep trying, but don't expect overnight miracles* (see Chapter 11 for a further discussion of changing attitudes).

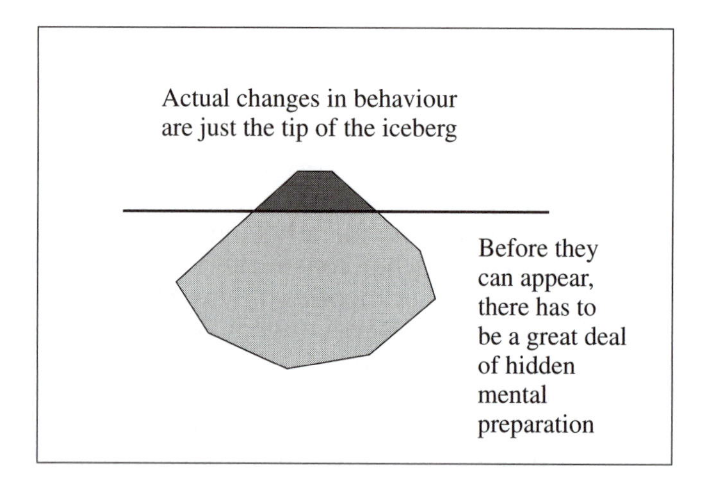

Figure 8.2 The ratio of behavioural changes to the underlying mental changes
Mental changes are not visible, but they are the necessary precursors of behavioural change

Starting Before you Begin

There is a great deal of evidence to show that achieving the right degree of mental readiness for a task before it actually begins can greatly enhance both the quality and the reliability of human performance. This evidence comes from recent psychological research and from studies of performance excellence, particularly Olympic athletes and top surgeons.[6] Both types of study show that considerable performance benefits can be gained by working carefully through a job *in imagination* beforehand. Psychologists call this *mental rehearsal*. The ability to acquire and practise these mental skills is what distinguishes the excellent performer from the average one.

In what follows, we will first describe the nature of this mental preparedness and then explain how it works to improve attentional focus. As stated earlier, error management is not about making bad people good, it is about making good people excellent.

Positive imagery
Prior to actually doing the job, top performers spend time thinking about the task and imagining how things will look and feel before they actually encounter them. Surgeons, for example, will convert two-dimensional textbook pictures into three-dimensional images of reality. They mentally rehearse the procedural steps in order to de-

velop a more fluid movement. The imagery is positive because they concentrate on visualizing the perfect result at each stage.

Being prepared for problems

Another important function of mental rehearsal is the anticipation of problems and preparing effective measures to deal with them. Top performers recognize the importance of being one step ahead of the task at all times. The following quote from a top surgeon catches the essence:

> I rehearse the operation in my head, and then sleep on it. It's very visual, step-by-step. I see myself doing it. I go through the different steps ... You think of the problems that could occur and what would be the solution. You can't always think of all the complications, but you have to be ready if they come up.

Mental readiness

This involves 'psyching yourself up' to do a good job. Surgeons, for example, do this by using study aids like books or models. They plan the procedural steps both in isolation and in consultation with colleagues. Many top surgeons do their best 'imagining' while exercising. An eminent cardiothoracic surgeon of James Reason's acquaintance swims for an hour each morning, a boring but fitness-inducing activity that he occupies by thinking about the difficult coronary artery anatomies he might encounter during that morning's operations.

Distraction control

Maintenance work is subject to interruptions and distractions. Recent psychological research has shown that the best way to ward off the effects of distractions is to anticipate them and to deal with each one individually as it occurs.[7] One study compared the performance of two groups, both engaged in solving arithmetic problems on a computer whilst subjected to intermittent distracting images at the top of the screen. One group was told to develop the mental 'set' of ignoring the images (disruption inhibition), the other was instructed to deal with the distractions by telling themselves they will do the arithmetic task as well as they could despite the distractions (task facilitation). The scores for the former group (the disruption-inhibiting group) were significantly higher than those for the task-facilitating group. However, both groups performed better than the controls who used no prepared distraction control method. The moral is clear: anticipate (as far as possible) the kinds of distractions and interruptions that are likely to occur and then devise—in advance—a specific strategy for dealing with each one. One such counter-measure is described below.

Avoiding place-losing errors
Maintenance work, in particular, is especially prone to the problem of losing or forgetting one's place in the sequence of task steps. Interruptions often lead to the omission of some necessary step after the job has been resumed (see Chapter 4). These errors could be minimized if the maintainer marks his or her place in the task (with a label or some other clearly visible indicator) at the time of being interrupted. These methods are most effective if prepared or thought about in advance.

Why is mental rehearsal so effective in creating reliable performance? How does it improve concentration and attentional focus? To answer these questions we need to return to the torch beam model of attention, presented earlier in Chapter 3 (see Figure 3.2, p. 23).

When we are carrying out a maintenance job, our performance is guided by three things: the written procedure (the work card or page of the manual), the reality in front of us and a 'mental model' of what should happen next. This mental model is made up of task-related images and knowledge structures (schemata) that we carry around in long-term memory. It is often incomplete or inaccurate, but we need it because we cannot (do not) always read the procedures and perform the job at the same time. The benefits of mental rehearsal lie primarily in improving the quality of the mental model and in activating the correct pathways (task steps) towards our intended goal. This gives them a better chance of claiming the attentional torch beam at the appropriate time. In other words, the right steps rather than the wrong ones are more likely to win out in the competition to capture attentional focus. These ideas are illustrated in Figure 8.3.

In Figure 8.3, the rectangle (seen in perspective) is equivalent to the bottom part of Figure 3.2 except that it is located in the mind rather than in reality. The circles represent various aspects of the equipment-to-be-maintained. The dark circles with the arrow running through them constitute the correct task steps needed to achieve

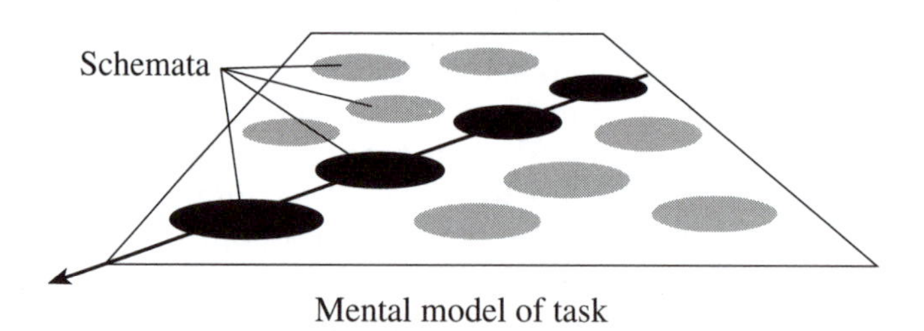

Figure 8.3 Primed schemata (knowledge structures) located within a mental model of the task

the desired goal. It is this internalized model of the task that is worked upon during mental rehearsal. The principle is simple. The more often these steps are activated, either in imagination or in actual performance, the more likely they are to attract the appropriate amount of attentional focus, and to ward off distractions.

Another way to think about Figure 8.3 is to imagine a board covered with a matrix of lights. Each light represents a possible action in respect to a piece of equipment. A light gets steadily brighter every time an action is performed, either in reality or in imagination. That is the important point: mentally picturing an action has the same effect as actually carrying it out. This not only makes the person more sensitive to the appropriate perceptual cues, it also strengthens the linkages between the correct actions. Furthermore, the brighter the light, the more likely it is to attract attention to it. Mentally rehearsing the correct sequence of steps creates a brightly lit pathway through the matrix of possible actions. The intensity of the lights will act to draw attention along the desired route and make it less vulnerable to distractions and interruptions.

Team Measures

Most of the system accidents discussed in Chapter 6 involved team errors of one kind or another. A team is a number of people working together to fulfil some common purpose or task. A large part of maintenance work is carried out in teams, particularly in the aircraft and nuclear power industries. Working in teams is a two-edged business. The need for several people—both within teams and between teams—to share the same awareness of what is happening or has happened can create increased opportunities for error. Yet the presence of fellow team members can also create an increased likelihood of error detection and recovery.

Commercial aviation has long been very good at instilling and certifying the necessary technical skills. But only within the past decade or so has the industry begun systematically to address the requirements of the human factors contribution to quality, reliability and operational effectiveness.[8] In analysing their accidents, the aviators have concluded that team management errors are one of the most serious threats to safety. These problems include:

- team leaders being over-preoccupied with minor technical problems
- failure to delegate tasks and responsibilities
- failure to set priorities
- inadequate monitoring and supervision

- failures in communication
- failure to detect and/or challenge non-compliance with standard operating procedures
- excessively authoritarian leadership styles
- junior members of the crew or team being unwilling to correct the errors of their seniors.

All of these factors combined in 1977 to cause the world's worst aviation accident, when two jumbo jets collided on a fog-bound runway in Tenerife with the loss of over 500 lives. The accident occurred when the captain of the KLM aircraft attempted to takeoff, wrongly believing that he had a takeoff clearance, and unaware that a PAN AM aircraft was taxying along the runway. We have made the point previously that sometimes the best people make the worst errors. In this case, the captain of the KLM aircraft was one of the airline's most senior and respected pilots, and was head of the KLM flight training department. A host of human factors contributed to this disaster, including communication failures, time pressures and an apparent failure by the copilot to assertively question the captain's decision to takeoff.

This terrible accident marked a turning point for the aviation industry. The Tenerife accident and others that followed in the late 1970s and 1980s prompted a major rethink of safety training, as airlines trained pilots to deal with the sort of team management issues listed above. Such training, commonly known as Crew Resource Management (CRM) training, is now provided by most airlines around the world, and is a legal requirement in most of Europe and North America. The success of these programmes has led to its extension beyond the cockpit. Increasingly airlines are introducing CRM training for maintenance personnel, sometimes known as Maintenance Resource Management (MRM).[9]

The goals of maintenance CRM courses include the following.

- Teaching team members how to pool their intellectual resources.
- Learning to acquire a collective situational awareness that admits challenges from junior team members.
- Emphasizing the importance of teamwork.
- Establishing a common terminology to minimize communication problems.
- Training for leadership and team membership skills.
- Identifying organizational norms and their effects on safety.
- Understanding organizational culture and recognition of shared values.
- Improving communication skills.
- Understanding and managing stress.

In applying CRM to aircraft maintenance, Continental Airlines uses two-day seminars, with 20–25 participants and two facilitators.[10] The first day includes a case study of poor teamwork and another illustrating the distinction between perception and reality, discussions of different behavioural and decision-making styles, assertiveness attributes and various stress management factors. The second day covers three major areas: teamwork and decision-making activities, case studies on recognizing organizational norms and interactive discussions of interpersonal skills (listening, supporting, confronting and feedback). The training programme was initially set up for managers and then extended to aircraft mechanics.

Northwest Airlines, another pioneer in applying CRM techniques to maintenance, has four components to its training (which it terms Maintenance Resource Management [MRM]). These are:

- communication skills
- crew development and leadership skills
- workload management
- technical proficiency.

Comparisons pre- and post-CRM courses showed a significant decline in ground damage incidents and improvement in dispatch reliability. Attitude and behavioural surveys before and after CRM courses also revealed a strong positive effect.

There is extensive evidence to show that CRM for maintainers has definite and measurable pay-offs not only in safety, but also in terms of improved reliability and cost reductions. A recent study looked at the impact of maintenance-related CRM courses in 150 US aircraft repair centres.[11] The main findings were:

- enthusiasm for the potential value of CRM-type training is high immediately after attending a course
- attitudes reflecting positive CRM values increase by 15–20 per cent immediately after training
- CRM training leads to lower occupational injury rates and a decrease in aircraft ground damage incidents. This is also associated with high aircraft dispatch rates.

There is no one right way to run maintenance CRM courses. 'Off-the-shelf' training is unlikely to be as effective as a course tailored to the specific needs and culture of your own organization. Nevertheless, there are some good examples of maintenance CRM courses that are freely available. The Federal Aviation Authority (FAA) has published a draft MRM Advisory Circular that contains a sample curriculum for such a course. The UK CAA has also produced a course in maintenance human factors that can be adapted to the needs of any business.[12]

In many major airlines, CRM training for aircrew has been in place for some 10–15 years. Although few would dispute its long-term value, these courses have not always been well received. As a result of these hard lessons, we now know something about how not to run CRM courses.

An article entitled 'How to kill off a good CRM programme' identifies the top 10 ways in which such schemes can fail.[13] They are listed below:

- Not integrating CRM into other forms of operational training (CRM, by itself, does not provide technical skills; if it is to be effective in operational terms, it must be integrated systematically into other forms of training).
- Failing to recognize the unique needs of your own company's culture.
- Allowing CRM zealots to run the show.
- Bypassing the research and data-gathering steps.
- Ignoring those concerned with training and checking standards.
- Having lots of diagrams, boxes and acronyms.
- Making the CRM programme a one-shot affair.
- Using pop psychology and psychobabble.
- Turning CRM into a therapy session.
- Redefining the 'C' in CRM to mean 'Charismatic' (where participants remember more about the instructor and how entertaining he was than what has to be applied in the operational setting).

The history of CRM in the aviation industry has been marked by a number of pendulum swings. The items listed above represent the main factors causing 'anti' feelings. The antidote for keeping any CRM-type programme on track is relatively straightforward.

- Operational management must stay involved with the programme as active supervisors and participants.
- Those possessing high operational credibility should give the instruction (that is, maintenance personnel themselves, not outside specialists or consultants).
- CRM must be fully integrated into training activities.
- It is essential to create a feedback system for tracking trends and participants' reactions, and for monitoring quality.
- CRM training is not a one-shot immunization against maintenance error. Regular reinforcement and refresher training is essential.

Summary

In this chapter we have examined some practical error management techniques directed at the level of the individual maintainer and the work team. We considered how with an awareness of fundamental human performance limitations, maintainers can learn to recognize danger signs which may signal an increased chance of error. We revisited the issue of violations and saw how attitude change can help to reduce rule-breaking behaviour, and we considered how mental preparation can increase the reliability of human performance. Finally, Crew Resource Management training was discussed. We considered the essentials of effective CRM courses, and the ways in which CRM training can fail.

Despite having written a good deal about error management measures directed at the individual, we believe that a warning is in order. On their own, these person- and team-focused techniques are not enough. We cannot take people out of the system, give them human factors training, then return them to the same unreformed system and expect dramatic improvements in their performance. As we will see in the next two chapters, measures directed at individuals and teams are of limited use unless they are accompanied by error management interventions directed at the workplace, the tasks and the system as a whole.

Notes

1 ICAO, *Annex 1, 4.2.1.2. e.* (Montreal: International Civil Aviation Organization); JAA, JAR 66 AMC 66.25 module 9 (Hoofddorp, Netherlands: European Joint Aviation Authorities, 1998).

2 D. Schwartz, 'Maintaining operational integrity through the introduction of human factors training', *The CRM Advocate*, **93**, 1993, pp. 1–2.

3 See J. Reason, D. Parker and R. Free, *Bending the Rules: The Varieties, Origins and Management of Safety Violations* (Leiden: Faculty of Social Sciences, University of Leiden, 1994).

4 B. Brehmer, N.-P. Gregersen and B. Moren, *Group Methods in Safety Work* (Uppsala: Institute of Psychology, Uppsala University, 1991).

5 See C. Sansone and J. Harackiewicz (eds), *Intrinsic and Extrinsic Motivation* (San Diego, CA: Academic Press, 2000).

6 T. Orlick, *In Pursuit of Excellence* (3rd edn) (Ottawa: Zone of Excellence, 2000).

7 P.M. Gollwitzer and J.A. Bargh (eds), *The Psychology of Action* (New York: Guilford, 1996).

8 See E. Wiener, B. Kanki and R. Helmreich, *Crew Resource Management* (New York: Academic Press, 1993); N. Johnston, N. McDonald and R. Fuller (eds), *Aviation Psychology in Practice* (Aldershot: Avebury, 1994).

9 B. Sian, M. Robertson and J. Watson, *Maintenance Resource Management Handbook* (Washington, DC: Federal Aviation Administration, 1998) <http://hfskyway.faa.gov>.

10 W.B. Johnson. 'Human factors in maintenance: An emerging system requirement', *GroundEffects*, **2**, 1997, pp. 6–8.
11 J.C. Taylor, 'Evaluating the effectiveness of Maintenance Resource Management (MRM)', Paper given to 12th International Symposium on Human Factors in Aircraft Maintenance and Inspection, 11–12 March 1998, Gatwick.
12 FAA, Draft Maintenance Resource Management Advisory Circular (Washington, DC: Federal Aviation Administration, undated); W.J. Done, 'Safety, human factors and the role of the regulator', Paper given to 15th Symposium on Human Factors in Aviation Maintenance, 27–29 March 2001, London.
13 W.R. Taggart, 'How to kill off a good CRM program', *The CRM Advocate*, **93**, 1993, pp. 11–12.

9 Workplace and Task Measures

The error management strategies presented in the previous chapter were directed at individuals and teams—improving skills, changing attitudes and beliefs or addressing coordination and communication. While making good people better is a worthy aim, it will have only limited effect if the work environment is one that continues to provoke errors, and particularly if the same error types keep recurring.

Because many maintenance errors have their origins in the work environment, some of the most powerful interventions to reduce error are those directed at removing task-related challenges to work quality. In this chapter, we discuss several environmental and task factors known to provoke errors and violations. These are fatigue, task frequency, design issues, housekeeping and spares, tools and equipment.

In the latter part of the chapter, we look in detail at task-oriented measures designed to reduce the incidence of omissions—the commonest form of maintenance error. The management of omissions involves two steps: first, the identification of omission-prone steps in a task; second, the provision of suitable reminders.

Fatigue Management

As shown in Chapter 5, fatigue can increase the likelihood of error in much the same way that alcohol does. No organization would encourage intoxicated workers to report for duty, yet many organizations will allow fatigued workers to do so, even though the risks may be similar.

If maintenance work is carried out outside standard hours, then fatigue management is one of the most important issues facing the organization. Well-designed rosters are one way to reduce the risks of fatigue. Eight principles for the design of shift systems are listed below.[1]

1 There should not be more than three night shifts in succession.
2 Avoid permanent night work.
3 Rotate shifts forward in time (morning → evening → night).
4 Allow at least two days off after the last night shift.
5 Consecutive work time should not exceed between five to seven days.
6 When shifts are longer than eight hours, no overtime should be permitted.
7 Allow at least 11 rest hours between shifts.
8 Where possible, ensure that shift schedules are known well in advance, and restrict subsequent modifications to a minimum.

While it was once difficult to evaluate shift rosters for work-related fatigue, there are now several computer programs designed to assess the level of fatigue that a shift pattern is likely to produce. A widely used program is that developed by Drew Dawson and Adam Fletcher at the Centre for Sleep Research in Adelaide.[2] Their program assigns a fatigue score for each period during the work shift.

A score of 40 would be typical when the working week is confined to Monday to Friday 09.00–17.00 hrs. The fatigue score will increase, however, with longer work periods, shorter break periods and/or break periods that do not allow for night-time sleep. According to Dawson and Fletcher, a fatigue score of 80 can lead to performance impairments similar to those obtained with a blood alcohol concentration of 0.05 per cent. For this reason, they recommend that workers performing safety-critical functions should not have fatigue scores above 80. Nevertheless, in some industries, fatigue scores greater than 150 have been obtained for individual workers.

As an illustration of Dawson and Fletcher's system, Figure 9.1 shows the predicted fatigue scores for a worker on a rotating pattern of morning, afternoon and night shifts, with a two-day break between each five-day period of work. As can be seen, the last hours of the fourth and fifth night-shift are producing fatigue scores greater than 80, suggesting that a change to the shift pattern may be necessary.

A major advantage of such fatigue management software is that managers can evaluate and compare possible shift patterns even before they have been worked.

Task Frequency

Engineers once used the familiar 'bathtub curve' to describe the frequency of failures throughout the life of a piece of equipment. According to this model, now somewhat out of favour, the early

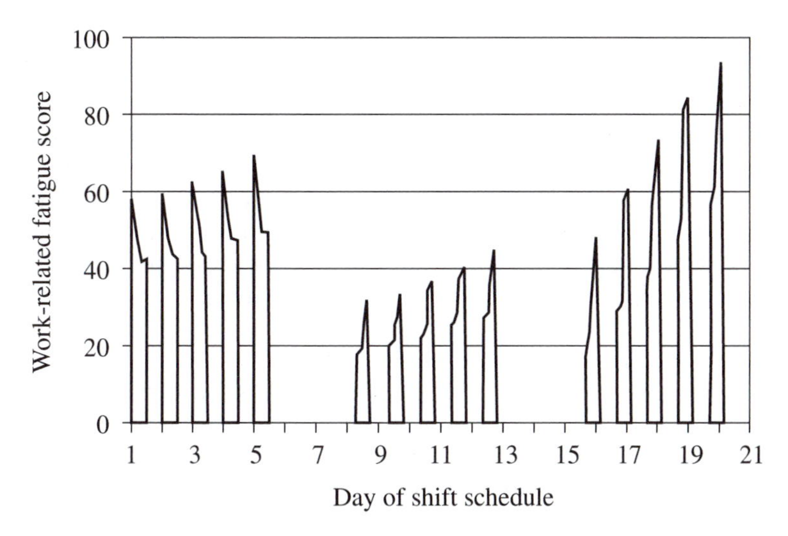

Figure 9.1 Relative work-related fatigue predictions for a five on (06.00 to 14.00 hrs), two off, five on (14.00 to 22.00 hrs), two off, five on (22.00 to 06.00 hrs), two off cyclic roster

Source: A. Fletcher and D. Dawson, 'A quantitative model of work-related fatigue: empirical evaluations', *Ergonomics*, **44**, 2001, pp. 475–88.

bedding-in period tends to have a relatively high risk of failure, but once these initial problems are resolved, a period of higher reliability usually follows, and then later, as the equipment nears the end of its design life, failures become more frequent. The reliability of maintenance performance can show trends that relate not to equipment life cycles, but to the experience of the maintenance personnel. The error rate for tasks that are performed infrequently is likely to be high, as inexperienced workers will tend to perform at the error-prone knowledge-based level. Once experience has been gained, however, the likelihood of knowledge-based errors diminishes, but the probability of skill-based slips and lapses increases. Absent-mindedness, as noted in Chapter 4, is a problem for the expert rather than the novice.

In assigning tasks to maintenance personnel, it is worth considering where along this continuum of experience the worker will approach the task. Both unusual and routine tasks can create their own kind of errors; however, intelligent task assignment can help to reduce these risks.

Design

Many maintenance errors have their origins in inadequate system design. Most maintainers can list examples of components that can be installed upside down or back to front, or systems that are difficult to access, or tasks that have apparently been designed with three-handed maintenance personnel in mind. Ease of maintainability has generally been a low priority for system designers. Six design principles for the maintainability of systems are listed below.[3]

1 There should be easy access to components.
2 Components that are functionally related should be grouped together.
3 Components should be labeled clearly and informatively.
4 There should be minimal need for special tools.
5 It should not be necessary to make delicate adjustments in the field.
6 Equipment should be designed to facilitate fault isolation.

Modern maintenance activities require the use of a wide variety of sophisticated testing, measuring and diagnostic equipment. Each item comes with a user interface upon which the maintainer acts to achieve some goal and from which he or she receives information. Both of these stages—execution (acting) and evaluation (information)—can provoke error and misunderstanding through poor design. This usually arises from a failure on the part of the designer to appreciate the user's perspective. Don Norman, a distinguished American psychologist, has identified two principal ways in which misunderstandings can arise: the *gulf of execution* and the *gulf of evaluation*.[4] These two basic design problems are shown in Figure 9.2.

Many problems arise because the designer assumes that their model of the system and the user's model are one and the same. But this is not necessarily the case since designers rarely talk directly to users, nor are they always well informed about the varieties of possible human errors. The user's model is derived from the system image—made up of its documentation together with what the user presumes the equipment is designed to do. If this does not make the designer's model clear, then the user will have an erroneous perception of the system's function.

One way to avoid these problems is to ask user-centred design questions before purchasing the equipment—these questions should be posed to a group of potential users. The first and most fundamental issue is how easily can the user determine the function of the system or device? Then there are three questions relating to execution.

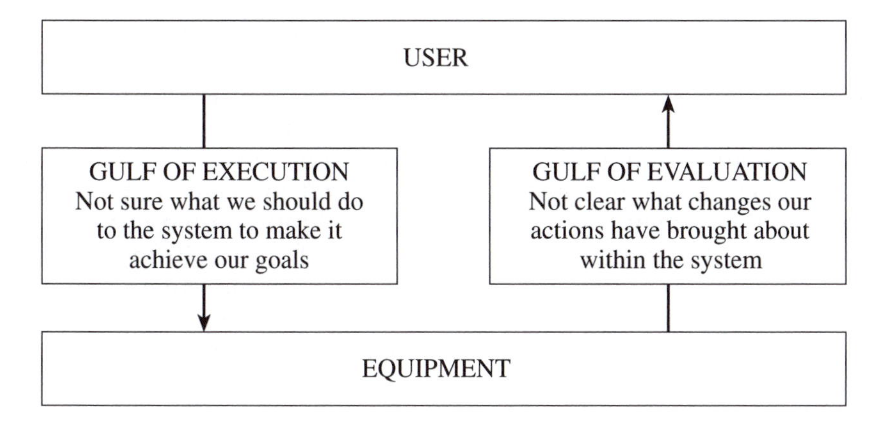

Figure 9.2 The gulfs of execution and evaluation between the user's model of what is going on and the actual state of the system

Source: After D.A. Norman, *The Psychology of Everyday Things* (New York: Basic Books, 1988).

- *Intention*. How easily can the users tell what actions are possible?
- *Action specification*. How easily can the users determine the mapping from intention to action?
- *Execution*. How easily can the users perform the actions?

Then there are a further three questions relating to evaluating the state of the device or system.

- *Perception*. How easily can the users tell what state the system is in?
- *Interpretation*.How easily can users determine the mapping from system state to correct interpretation?
- *Evaluation*. How easily can users tell if the system or device is in the desired state?

While it is not possible to design systems that eliminate the possibility of user errors altogether, both designers and purchasers can do much to reduce their occurrence. For example, one of the common characteristics of modern electronic devices is that they usually perform more operations than they possess dedicated keys or controls. To obtain the full functionality of the equipment, the user must first select the desired mode of operation before using the limited key or control set. This paves the way for one of the commonest error types that arise when using such equipment. They are *mode errors* that involve performing the operations appropriate for one mode when in fact the user is in another mode, and does not realize it. Such mode errors have been responsible for a number of air accidents, where flight crew have entered manoeuvring instructions into the flight management system while erroneously believing it to be in another mode. One of the problems with such flight management systems was that the current flight mode was poorly represented to the user. It is thus vital that the potential purchasers of any multi-modal item of maintenance equipment should establish at the outset that the information given to the user about the current mode is clear and unambiguous. In short, ensure that the modes are distinctively marked and easy to interpret.

Housekeeping

Issues such as housekeeping practices are powerful indicators of the culture of the organization. If materials such as wire off-cuts are left lying on the floor after a job is completed, or removed components are stored in a haphazard way, then the environment is likely to be one in which errors are more frequent, and more severe in their consequences.

Poor housekeeping constitutes a clear sign of system malaise when it has been present for a long time and nothing has been done about it. Poor housekeeping cannot go unchecked for long periods unless there are management failures. These failures can be of three kinds.

- Management carries out inspections, is aware of the problem, but does nothing about it.
- Management makes inspections, but is fooled and is thus unaware of the problem.
- Management does not make inspections often enough.

The trick with housekeeping is to tread a middle path that, on the one hand, avoids the extremes of an over-punctilious concern with cleanliness, tidiness and outward form, and, on the other, avoids neglecting dangerous slovenliness. Either extreme carries a penalty. It is necessary to find a standard of housekeeping that meets the needs of safe, swift and effective operations, but does not go too far beyond these aims.

Spares, Tools and Equipment

Practical issues such as the availability of spares and equipment can be powerful initiators of human error as workers struggle to perform their tasks in the face of obstacles and frustrations. No special knowledge of psychology is required to manage these issues, just a recognition that an important part of managing error is getting the task environment right. These issues could form the core of an audit of the workplace. Some key questions regarding tools and equipment are listed below.

- Do workers ever use unapproved tools or equipment?
- Is unserviceable maintenance equipment left in work areas while awaiting repair?
- Are commonly used spares or consumable items out of stock?
- If maintenance is carried out around the clock, is technical support available at all hours?
- Are there systems in place to keep track of tools?
- Are work areas tidy?
- Are disassembled components stored and labelled in appropriate ways?
- Are replacements available when equipment is sent away for servicing or calibration?

Using Procedures to Manage Omissions

In Chapter 2, we identified the omission of necessary steps during reassembly or installation as the most common form of maintenance error. Some estimates suggested that omissions accounted for more than half of all the human factors problems encountered in maintenance work. In the remainder of this chapter, we will focus upon the management of omission errors through the prior identification of omission-provoking task steps in a maintenance procedure and the provision of suitable reminders.

Omissions are particularly dangerous since they can create common mode failures in which a number of functionally related components lying 'downstream' of the missing part or substance can be rendered unserviceable. Moreover, omissions may lie concealed for some time before discovery or, worse, before they interact with local triggering events to cause an accident or an incident. Thus, a strong case exists for treating omissions as a specially targeted category for error management. Fortunately, we know a good deal about the characteristics of task steps that are likely to provoke omissions. Knowing in advance where an omission is likely to happen is at least halfway towards effective EM. The other half is finding effective ways of drawing people's attention to the possibility of omission so that they might avoid it.

Omission-provoking Features

Four factors, in particular, determine the omission-likelihood of a task step.[5] The first is the *memory loading* of the step: that is, the amount of information that has to be carried in a person's head in order to complete the step. The higher the memory loading of a step, the more likely it is that some or all of it will be omitted. The second factor is *conspicuity*. Items that are concealed or inconspicuous are more likely to be omitted, especially during reassembly when replacing them requires a distinct act of memory. The third factor is the step's *position* within a task sequence. Two positions are vulnerable: mid-sequence steps and those toward the end of a sequence. The vulnerability of mid-sequence steps depends upon their nature, but nearly all steps that occur towards the end of a task are particularly prone to forgetting or premature exits. The fourth factor is *local cueing*. Many steps are cued or prompted by the preceding one. For example, a bare bolt prompts the replacement of washers and nuts, an empty screw hole cues the insertion of a screw and so on. Often, however, steps are functionally isolated from the rest of the task, particularly during reassembly. This isolation renders them especially vulnerable to omission.

All of these factors combine to create one of the commonest omissions in everyday life: that is, leaving the last page of the original behind after having photocopied a loose leaf document on a simple photocopier. First, the act of removing the last page of the original is functionally isolated. The removal of the previous pages is prompted by the placing of each new original. Second, the last page is concealed under the lid of the photocopier. Third, it occurs right at the end of the photocopying sequence when one's mind is already on the next task. Fourth, the end position is further complicated by the fact that it occurs after the main goal of the task has been completed—we have just seen the last page of the copy emerge, something that acts as a powerful signal for departure. The omission-provoking effect is additive. The more features that combine within a single step, the more likely it is to be omitted.

A Task Step Checklist

One way of converting these general principles into a practical error management tool is to use the 20-item checklist set out in Table 9.1.

The Task Step Checklist is used in relation to a specific task procedure by someone having experience of the job and with access to a quality lapse or error reporting system (someone in quality assurance, for example). The procedure set out below happens to be

Table 9.1 Task Step Checklist

INSTRUCTIONS: Apply this checklist to all the task steps identified in a task analysis or manual. If in doubt, score the feature affirmatively. Sum the scores for each task step and enter into the total column in the scoring grid [see text discussion].

	OMISSION-PRONE FEATURE	SCORE
1	Has this step ever been omitted in error in the past?	If yes, score 3
2	Does this step form part of an installation or reassembly sequence?	If yes, score 3
3	Does this step involve routine and highly practised actions?	If yes, score 2
4	Does this step involve consulting written procedures that do not always correspond with local conditions?	If yes, score 2
5	Is this step functionally isolated from the rest of the sequence (i.e., not obviously cued by preceding actions, or stands apart)?	If yes, score 2
6	Does the performance of this step involve a recent change from previous practice in carrying out this task?	If yes, score 2
7	Does this step involve actions or items not required in other very similar tasks?	If yes, score 2
8	If this step were omitted in error, would its absence later be concealed from view?	If yes, score 2
9	Does this step involve the repetition of actions (i.e., recursions) that depend upon some local condition or requirement?	If yes, score 1
10	Does this step involve installing multiple items (bushes, washers, nuts, etc.)?	If yes, score 1
11	Does the step require cues, or items that are not easily visible, detectable or readily to hand?	If yes, score 1
12	Does this step occur near to the end of the task?	If yes, score 1
13	Does this step occur after the achievement of the task's main goal but before its actual completion?	If yes, score 1
14	Is the performance of this step liable to interruptions or external distractions?	If yes, score 1
15	Is this step likely to be carried out by someone who did not start the task?	If yes, score 1
16	Is the performance of this step conditional upon some earlier action, condition or state?	If yes, score 1
17	Does this step require remembering detailed instructions?	If yes, score 1
18	Does this step require the removal of tools or unwanted objects from the immediate location of the task?	If yes, score 1
19	Does this step involve the installation and adjustment of multiple fastenings?	If yes, score 1
20	Is this step sometimes not required during the execution of the present task?	If yes, score 1

described in relation to the changing of an aircraft wheel, but any other maintenance task would serve as well:

1 Select a task that has omission-prone actions or items, particularly one in which omissions could seriously jeopardize safety.
2 Break the task down into its component steps. Each step is defined as a set of actions that achieves a necessary sub-goal (for example, install the axle washer). Do not attempt to break the step down into its constituent actions (for example, find axle washer, grasp axle washer, place axle washer on nose gear axle and so on). This is not a time-and-motion study. The level of detail provided in, say, a Boeing manual is adequate to label the step.
3 Enter a summary description of the task steps into a score sheet. The score sheet should have the task steps as rows and the omission-provoking features as columns. Any one step may score on several features, so the score sheet should include 20 scoring columns, one for each of the error-provoking features.
4 It may be judged that certain steps are so unlikely to be omitted (or that such an omission would be so quickly discovered) that they should not be included in the list of steps. For the removal and installation of an aircraft's nose gear wheel and tyre, examples of these unnecessary-to-include steps might be *remove the axle protector* or *use a wheel change dolly to move the wheel and tyre assembly to its position on the axle.*
5 Use the Task Step Checklist to score every identified step for each of the 20 omission-prone features (see Table 9.1). Note that these features carry different scores reflecting their judged importance in predicting omissions. Use professional engineering judgement in deciding upon 'yes' and 'no' responses for each feature. If there is any doubt, then it is better to score the item as a 'yes'.
6 Identify the 5–10 highest scoring steps. These scores are obtained by summing across the features for each step and entering this number into the total column. Consider preparing suitable reminders for each of these task steps. The characteristics of a good reminder are discussed on pp. 130–32.

Additional Explanatory Notes

Further explanation of each of the 20 omission-provoking features is given in the notes listed below. Each note relates to the same numbered item in the Task Step Checklist (see Table 9.1).

1 Previous omissions of this step may or not have been recorded as quality lapses. It is important to find out from those who

regularly perform this task whether there have been unrecorded (or later detected and recovered) instances of this step being omitted.

2. The important distinction here is between disassembly and reassembly. Putting things back is many times more prone to error than taking them apart.
3. Absent-minded omissions are only likely to occur in tasks or sub-tasks that can be run on 'automatic pilot'.
4. Procedures, manuals or work cards in which the text or the diagrams do not match up to the reality as seen by the person doing the job are frequently implicated in erroneous omissions.
5. This is a feature that requires a certain degree of judgement on the part of the analyst. A step is functionally isolated if it stands apart in some way from the rest of the task sequence, and/or if it is not obviously cued by preceding actions.
6. This feature applies when there has been a recent change in carrying out this task. It may have been decided, for example, that an extra inspection or check is now required at a particular stage in the task sequence. Or that some other additional actions are now required. These 'add-ons' are easily forgotten.
7. This feature involves negative transfer, or the case where a person moves from one comparable task to another in which there are many similarities but also some important differences. On some aircraft, for example, the wheel spacer is attached to the wheel, on others it is not. This has contributed to omitting to install the wheel spacer in the latter case. This is a situation that is likely to recur when technicians are moving from one aircraft type to another, and where there are often only subtle differences between the actions required.
8. Many omissions, like missing washers, spacers, caps, fastenings and so on, are often concealed from view by subsequent reassembly or installation activity. This makes an omission much harder to detect and recover.
9. Steps involving recursions or the repetition of previous actions in order to satisfy some local condition are especially prone to omission.
10. It is often the case that a step involves the installation of more than one component (for example, three washers, 10 nuts and so on). Some of these items are liable to be left out.
11. A step or an item that is not conspicuous, or readily to hand, is subject to the 'out-of-sight-out-of-mind' principle and liable to be omitted.
12. Steps occurring near to the completion of the task are subject to premature exits in which an individual moves on to the next job without completing the first.

13 Sometimes the main goal of the task is achieved before all the necessary task steps are completed. For example, in a main gear wheel and tyre installation, the main goal—installing the wheel and replacing the hubcap assembly—is achieved before the task is finished. It still remains to close circuit breakers, remove tags, test tyre pressure and so on. These late steps are particularly vulnerable to omission in conditions of time pressure and high workload.

14 All steps are, in principle, liable to distractions or interruptions, but some are more vulnerable than others. Deciding whether or not this is the case for this step is a matter for judgement and local knowledge.

15 In many maintenance activities, those who begin a job do not always finish it. Where this applies, there is a greater risk of omission.

16 Sometimes the performance of a particular step depends upon some state or condition encountered earlier in the task. These conditional steps are readily forgotten.

17 Many steps in maintenance require keeping a large amount of information in memory or having procedures close to hand. Since people rarely 'read and do' at the same time, there is always a strong chance that some of the information required to perform the step correctly will be forgotten and hence omitted.

18 Evidence from quality lapse and error data shows very clearly that the need to remove tools and foreign objects from the job area is frequently forgotten and hence omitted. Again this suffers from occurring late on in the task sequence, but it is still a special and distinct case.

19 Fastenings, particularly multiple fastenings, are especially prone to omission. Again, they are subject to premature exits.

20 Steps that are required on some occasions but not others are liable to omission, particularly if the need to perform such steps occurs relatively infrequently.

The Characteristics of a Good Reminder

The people who actually do the job are usually the best judges of what sort of reminder they should use to mark an omission-prone step or item. Regardless of what form the reminder takes, however, there are a number of principles that can be used to create an effective memory jogger. These are listed in Table 9.2.

Most maintenance organizations already use reminders of one kind or another. It will be readily appreciated therefore that reminders, like any other single EM tool, have their limitations. People forget to use them or ignore them. They also have a diminishing impact. The

Table 9.2 Ten criteria for a good reminder

In order to work effectively, reminders (memory aids to prevent the omission of necessary task steps) should satisfy all of the five conditions described below.

CONSPICUOUS	A good reminder must be able to grab the maintainer's *attention* at the critical time.
CLOSE	A good reminder should be positioned *as closely as possible* in both time and distance to the location of the necessary task step.
CONTEXT	A good reminder should provide sufficient information about *when* and *where* the to-be-remembered task step must be executed.
CONTENT	A good reminder should provide sufficient information to tell the maintainer *what* has to be done.
COUNT	A good reminder should allow the maintainer to *count off* the number of discrete actions or items that need to be included in the correct performance of the task step.

In addition to satisfying these main criteria, a good reminder may also need to satisfy the secondary criteria set out below.

COMPREHENSIVE	A good reminder should work effectively for a *wide range* of to-be-remembered steps.
COMPEL	A good reminder should (when warranted or possible) compel the technician to carry out a necessary step by *blocking* further progress until it has been completed.
CONFIRM	A good reminder should help the technician to *check* that the necessary steps have been carried out as planned. In other words, it must continue to exist and be visible after the time for the step execution has passed.
CONVENIENT	A good reminder should not cause unwanted or additional *problems*, particularly if these turn out to be worse than the possible omission.
CONCLUDE	A good reminder should be *readily removable* once the time for the action and its checking has passed.

longer the same reminder is left in place, the more likely it is to become just another part of the background scenery. To remain effective, reminders have to be renewed and revitalized. But the effort is worth it. Even if they only succeed in preventing a quarter of all the possible omission errors, they will still have made a substantial impact upon a serious human factors problem.

Summary

In this chapter we have focused on several key aspects of the maintenance task and environment that have a powerful impact on error production. Error management is about getting the task and environment right as much as it is about focusing on human issues. We noted that principles are available to help guide the design of shift rosters and that fatigue evaluation software may assist in the design of shift systems. We then considered task frequency and system design, before briefly acknowledging the importance of fundamental issues such as spares availability, housekeeping and the availability of tools and equipment. Omissions are the most common form of maintenance error. Omission management requires particular attention, therefore, and we outlined an approach that can be used to identify omission-prone task steps. In the next chapter, we focus on error management at the widest level, that of the entire organization or system.

Notes

1 P. Knauth, 'Changing schedules: shiftwork', *Chronobiology International*, **14**, 1997, pp. 159–71.
2 D. Dawson and A. Fletcher, 'A quantitative model of work-related fatigue: Background and definition', *Ergonomics*, **44**, 2001, pp. 144–63; A. Fletcher and D. Dawson, 'A quantitative model of work-related fatigue: empirical evaluations', *Ergonomics*, **44**, 2001, pp. 475–88.
3 A.E. Majoris and E. Boyle, 'Maintainability', in G. Salvendy (ed.), *Handbook of Human Factors and Ergonomics* (2nd edn) (New York: Wiley, 1997). See also B.S. Dhillon, *Engineering Maintainability: How to Design for Reliability and Easy Maintenance* (Oxford: Butterworth-Heinemann, 1999).
4 See D.A. Norman, *The Psychology of Everyday Things* (New York: Basic Books, 1988). Also D.A. Norman and S.W. Draper, *User Centered System Design* (Hillsdale, NJ: Erlbaum, 1986).
5 J. Reason, 'How necessary steps in a task get omitted: Revising old ideas to combat a persistent problem', *Cognitive Technology*, **3**, 1998, pp. 24–32.

10 Organizational Measures

A recurrent theme of the previous chapters is that although errors are human actions, the conditions that promote them are a product of the maintenance system as a whole. At the risk of being overly repetitive, let us assert once more that managing error requires action not just at the level of the individual or the workplace, but at all the levels of the organization. This chapter discusses techniques for managing organizational factors known to exert a powerful 'upstream' influence on maintenance errors and maintenance mishaps.[1]

How Accidents Happen: A Reminder

Before we look at specific management tools, however, let us briefly review the nature of untoward events. An 'event', in this context, is any unwanted and unplanned happening. It can range from an inconsequential error to a catastrophic accident. All events have three basic components:

- *Causal ingredients*. Errors, violations, latent system conditions, technical failures and the like.
- *Timing*. The point in time at which the causal ingredients come together to create a pathway through some or all of the system's many defences.
- *Consequences*. As indicated above, these can cover a wide span from a trivial inconvenience to a disastrous loss of life and assets.

While we cannot easily control the timing of an event, we can remove some of the more damaging *causal ingredients* from the equation, thus making it less likely that the defences will be breached and that bad consequences will ensue. And we can do this in two ways: by using *reactive outcome measures* to learn the right lessons from past events, and by using *proactive process measures* to assess the system's 'safety health' and then deploying targeted remedial actions to increase its basic resistance to the operational hazards. By employing these two data-gathering techniques in a coordinated manner it is possible to identify and remove many of the ingredients of maintenance mishaps *before* they come together to cause injury, loss or damage. We will discuss both of these measures in the next section.

The *timing* of an event, particularly in complex, well-defended systems, is very largely in the hands of chance. We may have some influence over which cards are in the pack, but chance largely decides exactly when the cards will be dealt.

The *consequences* of bringing the various accident-producing ingredients together are shaped mainly by the local circumstances. If they are benign, then the outcome could be insignificant; if they are unforgiving, then the results may be catastrophic. These effects can be mitigated to some extent if management works hard to make the system as a whole more tolerant of mishaps.

Reactive and Proactive Measures: Working Together

The reactive outcome measures and the proactive process measures mentioned earlier function in a complementary way to identify which of the organization's core activities is most in need of improvement at any given time. The complementary use of these measures is summarized in Table 10.1.

All maintenance activities share a number of common factors that have a profound effect upon both commercial success and the 'safety health' of the overall system. At the organizational level, these factors include organizational structure, training and selection, people management, the provision of tools and equipment, commercial and operational pressures, planning and scheduling, communication and the maintenance of buildings and equipment. All of these practices, together with the largely unspoken beliefs, values and norms of the workforce and its management, act in concert to determine the organizational culture, of which the safety culture is a crucial component.

Such 'upstream' influences shape the local conditions within a particular workplace. These local factors include the knowledge, skills and ability of the workforce, the quality of tools and equipment, the availability of parts, paperwork, manuals and procedures, ease of access to the job, computer support and the like. These, in turn, have a direct influence upon the reliability and efficiency of individual maintainers. To return to the mosquito analogy first used in Chapter 2, they could be the 'swamps' in which errors and violations breed,

Table 10.1 How reactive outcome measures and proactive process measures can work together to reveal systemic and defensive weaknesses

	REACTIVE OUTCOME MEASURES	PROACTIVE PROCESS MEASURES
WORKPLACE AND ORGANIZATIONAL FACTORS	Analysis of many incidents can reveal recurrent patterns of cause and effect that are not perceptible in single events.	Regular sampling of the system's 'vital signs' reveals those most in need of correction. This leads to steady gains in 'fitness' or resilience.
DEFENCES, BARRIERS AND SAFEGUARDS	Each event shows a partial or complete penetration of the maintenance system's many layers of defence.	These regular checks also show where defensive weaknesses currently exist, and where they are likely to appear in the future.

or they could be the necessary conditions for sustained good performance. In either case, we need to establish a continuous cycle of monitoring and targeted improvements. If this review and reform process is neglected, then it is likely that these core activities will be the causal ingredients of future events.

Reactive Outcome Measures

Any company that is serious about reducing and containing human factors problems must first understand the nature and varieties of the errors that occur within its own system. A safe culture is an informed culture. It is not enough simply to collect and analyse those maintenance errors that have such bad consequences that they could not go unrecorded. These are only the visible tip of the iceberg. To appreciate the recurrent (and hence manageable) error patterns fully—and thus be able to deploy your limited EM resources in the most efficient manner—you need to know about the near misses and 'free lessons'. These are the errors that nearly happened, or that were detected and recovered without ill effects.

In order to obtain this knowledge, the workforce must be persuaded to report these low-cost events in sufficient detail and in sufficient numbers to be of use. This is no easy task. People do not readily confess their own blunders. But it can be made to happen, as several successful schemes have demonstrated. Two things are essential: trust and convenience of reporting. The issue of trust lies at the heart of a safe culture. Trust depends crucially on everyone understanding the difference between acceptable and unacceptable actions. This ratio is usually of the order of 90 : 10—that is, 90 per cent of maintenance-related errors fall into the blameless category and should incur no sanctions if reported. Indeed, their reporting should be rewarded. Chapter 11 discusses the ways in which a just culture (and hence a reporting culture) can be socially engineered. In many respects, this is the most important chapter in the book. Without a supportive safety culture, any attempts at error management are likely to have only very limited success.

Once a sufficient number of errors have been reported, it is possible to analyse them according to a variety of factors that may be associated with their occurrence. Such error factors could include people, tasks, locations or types of equipment.

Figure 10.1 depicts two possible distributions of error frequencies in relation to various error factors. The distribution on the left shows an even distribution of errors over factors—of the kind that would be expected if the relationship between the occurrence of errors and the factors on the horizontal axis were largely random. The distribution on the right, however, shows a significant clustering of errors in relation to specific factors, suggesting that only some of them are instrumental in provoking human factors problems. It is these systematic patterns that are of greatest interest because they permit the targeting of limited error management resources at those areas most demanding of further investigation and treatment.

The existence of recurrent error patterns provides an important clue to the existence of error traps within the system. Error traps are conditions within the workplace or system that keep producing the same kinds of error, regardless of who is doing the job. The problem

Figure 10.1 Distinguishing between random and systematic error patterns

clearly lies in the situation rather than the people. One of the defining characteristics of a safe organization is that it works hard to find and eliminate its error traps.

MEDA (Maintenance Error Decision Aid)[2] is a good example of an effective event analysis tool that was specifically designed for the maintenance environment. MEDA was the brainchild of David Marx (then a Boeing engineer) whose work on creating a just reporting culture in maintenance organizations will be discussed in the next chapter. Although many event-reporting systems exist within the maintenance world,[3] MEDA is one of the most principled and widely used tools. Boeing distributed it free of charge to all of its airline customers. Although tailored to the needs of aircraft maintainers, it can easily be adapted to all types of maintenance operations.

The MEDA results form is made up of five sections as summarized below:

- *Section I* gathers general information about the airline, the aircraft type, the time of the incident and so on.
- *Section II* describes the operational event (for example, flight delay, cancellation, gate return, in-flight engine shutdown, aircraft damage, injury, diversion and the like). The user is also required to give a free text description of the event in approximately 20 words or less.
- *Section III* identifies the nature of the maintenance error. Errors are classified under seven major headings:
  - Improper installation
  - Improper servicing
  - Improper fault isolation/inspection/testing
  - Foreign object damage
  - Surrounding equipment damage
  - Personal injury
  - Other.
- *Section IV* identifies the likely contributing factors existing within the workplace or the organization. These are described in detail below.
- *Section V-a* identifies the failed defences. That is, it asks whether there were any existing procedures, documentation, processes or policies that should have prevented the event, but did not.
- *Section V-b* requires the analyst to itemize suggested corrective actions that should be taken at both the local level and within the organization at large to prevent a recurrence of the event.

Thus, Sections I–III answers the 'what?' questions. Section IV answers the 'how?' and the 'why?' questions. Section V-a pinpoints failed system barriers, and Section V-b outlines potential solutions.

The essentially reactive information provided by MEDA and other tools for analysing past events can be augmented by more proactive measures that involve both the continuous assessment of known error-producing factors in the workplace and regular checks upon the systemic factors that contribute to an organization's 'safety health'.

Proactive Process Measures

Audits of one kind or another are perhaps the most commonly used proactive process assessment tool. Over the last two decades, however, other proactive devices have been devised that rely extensively on the standardized judgements of those at the 'sharp end'. Unlike audits, these tools are bottom-up devices, sending information upwards, rather than top-down instruments. They inform supervisors and managers about what work conditions are really like on the frontline. In particular, they serve to identify and prioritize those workplace and organizational factors that are creating an adverse effect on human performance. Process measures do not depend upon the prior occurrence of either errors or bad events; their purpose is to identify those workplace and organizational factors that may later cause an event, and to direct remedial efforts at those problems most in need of attention.

MESH (Managing Engineering Safety Health)[4] is one example of this type of proactive process measure. It was created for British Airways Engineering in the early 1990s, and later adapted for use by Singapore Airlines Engineering Company. MESH is a set of diagnostic instruments for making visible, within a particular maintenance workplace, the situational and organizational factors that are currently having the most adverse effects upon task performance. Collectively, these measures were designed to give an indication of the system's state of safety (and quality) health, both at the local workplace level and in general. For convenience, the whole diagnostic package was implemented within a linked suite of computer programs.

MESH was based upon the following assumptions.

- Safety is not just a matter of negative outcomes. It is a function of the system's intrinsic resistance to operational hazards and event-producing factors. This resilience is termed 'safety health'.
- Safety health is something that emerges from the interplay between many factors at both the local workplace level and the organizational level.
- A system's safety health can only be assessed and managed effectively through the regular measurement of a limited subset of these local and organizational factors.

- MESH is designed to provide the measurements necessary to sustain a long-term fitness programme.

Exactly what local factors are assessed depends upon the workplace. Some factors will be common to all locations, but others will vary from place to place. For example, the following 12 local factors were judged as being appropriate for a hangar in which line maintenance personnel carried out overnight 'first aid' repairs.

- Knowledge, skills and experience
- Morale
- Tools, equipment and parts
- Support
- Fatigue
- Pressure
- Time of day
- The environment
- Computers
- Paperwork, manuals and procedures
- Inconvenience
- Personnel safety features.

These factors were not intended to be a comprehensive listing. Rather, as in the way of opinion polls, they were meant to sample the quality of the work environment.

Assessments were made through simple subjective ratings of the extent to which each one of these local factors had been a problem in relation to a small number of jobs, days or tasks (how these were specified depended upon the particular work location). The assessments were made directly on a computer, using either the mouse cursor or the keyboard.

These local factor assessments were made by between 20 and 30 per cent of the 'hands on' workforce in any given location. The assessors were selected randomly, and they each made regular ratings. In some areas, these were made on a weekly basis, while in other areas monthly ratings were considered more appropriate. A given group of assessors operated for limited period, say a quarter, after which a new set of assessors was selected randomly, and so on. This method of selection is in keeping with the fact that MESH is a sampling tool.

The assessors were anonymous. On logging on to the MESH program, they were asked to give their grade, trade and location. On completing their ratings, the assessors were provided with a profiled summary of their own input together with a cumulated profile for all ratings made over the past four weeks.

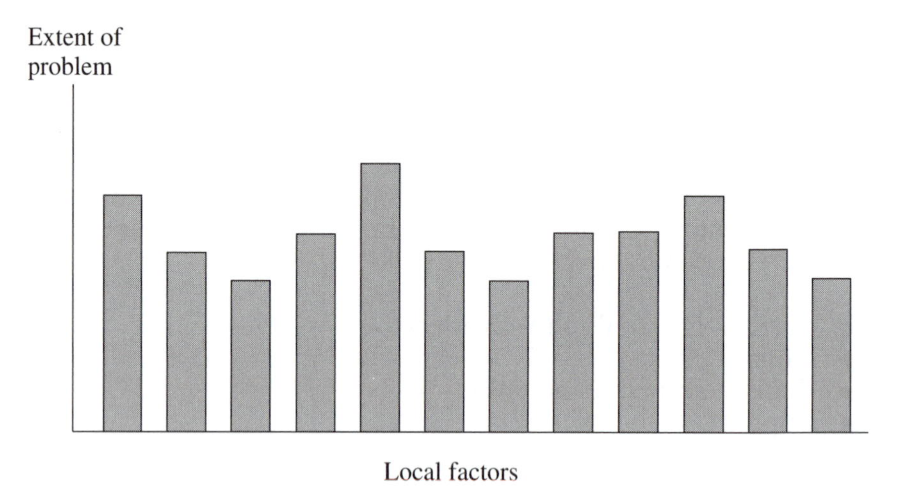

Figure 10.2 A local factor profile
Normally the specific local factors would be labelled on the x-axis

Figure 10.2 shows an example of a typical local factor profile. The y-axis indicates, on an ordinal scale, the relative extent to which each of the local factors had constituted a problem in carrying out a variety of tasks over a given time period. The profile shows, at a glance, which local factors are most in need of remediation. The intention was that the local management should focus on the worst two or three local factors only. Resources are always limited. It is much better to target specific problems than to try to tackle all of them. Thus, the local factor profile allowed management to prioritize their quality and safety goals for the upcoming period.

The MESH program also included a 'comments' facility that allowed users to describe specific instances of problems. This proved very useful in guiding the subsequent remedial actions. It should be stressed that the MESH comments facility supplemented but did not replace existing quality issue reporting systems.

Whereas the local factors to be assessed varied from one workplace to another, the same eight organizational factors are measured in each maintenance location. These are listed below.

- Organizational structure
- People management
- Provision and quality of tools and equipment
- Training and selection
- Commercial and operational pressures

- Planning and scheduling
- Maintenance of buildings and equipment
- Communication.

While frontline maintainers assessed the local factors, organizational factors needed to be judged by technical management grades—that is, people on the interface between the organization at large and their specific workplace. Since organizational factors are likely to change far more slowly than local factors, assessments were made at less frequent intervals, say monthly or even quarterly.

As with local factor assessments, the organizational factor data were summarized as bar chart profiles by the MESH program. The purpose of the profile was to identify the two or three organizational factors that were most in need of improvement. Subsequent profiles provided feedback as to the success or otherwise of these remedial measures.

In summary, MESH focused on the positive faces of quality and safety by making regular assessments of the system's 'vital signs' at both the local and the organizational levels. It provided 'soft' numbers for management to work with, and allowed them to prioritize human factors problems and to track progress in fixing them.

Proactive process instruments, of which MESH is but one example, are not just another safety 'add on'. They measure local and organizational factors that are as relevant to production and quality as they to safety. As such they comprise an essential part of any manager's professional tool bag. Nor do these tools replace existing safety and quality measures. They extend and unite them. Indeed, a prerequisite for the successful application of instruments like MESH is that these other 'good practices' are already in place. We will consider the integration of these measures further in Chapter 12.

Identifying Gaps in the Defences

Even if we do the best we can to prevent maintenance errors, we know that some will still occur. So we need to ensure that our systems are as error-tolerant as possible. We can do this by ensuring that we have appropriate defences in place. Two types of defences are important in this context: those designed to detect errors, and those intended to contain the consequences of undetected errors.

It is necessary to consider two issues. First, errors will happen. Second, it is possible—even likely—that these errors will have adverse consequences beyond the task in hand. Such thinking may not come naturally to maintenance personnel who are likely to be focused on their particular jobs. It therefore falls to those managers whose responsibilities extend beyond particular workplace boundaries to take active steps to improve the likelihood that inevitable errors will be detected and recovered. And, if this fails, to ensure that such errors do not have bad outcomes in some other part of the system. While individual errors are largely unpredictable, gaps in defences can be found at any time; but only if we go looking for them. Below is a list of questions that, if answered in the affirmative, suggest that there are weaknesses in the error detection defences.

- Is work self-inspected, or inspected by members of the same work group?
- Are functional checks ever omitted or abbreviated due to time pressure?
- Is insufficient time allowed for maintenance personnel to perform all the required functional checks?
- Are some clearly unnecessary functional checks required?
- Does the climate of the workplace discourage thorough checking of colleagues' work?
- Do some safety-critical tasks lack error-detecting defences?
- Are functional checks performed predominately at the end of a shift when personnel are likely to be fatigued?
- Have jobs ever been signed-off as completed and satisfactory when later events showed that this was not the case?
- Could a system pass a post-maintenance test, but then fail to work when returned to service?

If any of the following questions are answered in the negative, then there is a strong possibility that the error-containment defences are less than adequate.

- If maintenance is being performed on a 'live' system, is it carried out at a time that will cause least disruption to other parts of the system?
- Is the permit-to-work system adequate?
- Are strenuous efforts made to avoid the simultaneous disruption of multiple redundant systems?
- Is staggered maintenance used to avoid disruption by maintenance error?
- After maintenance, is the system operated in a forgiving environment before being returned to full service?
- Are operators or production personnel kept informed of recent maintenance activities?

We can never guarantee total immunity from bad events, but we can increase the system's intrinsic resistance by strengthening defences and by identifying and eliminating (as far as possible) the known causal ingredients that are latent in all systems. Zero accidents, no mishaps and a total freedom from human performance problems are not realistic goals. But being as resistant as reasonably practicable to your operational hazards is an achievable aim. Other than closing down your facility completely, this is the best you can hope for and still stay in business.

Summary

The chapter began by reminding the reader of how events happen. There are three components: the causal ingredients, the timing of their interaction and the consequences. While we can do little to affect the timing, or avoid errors altogether, we can seek to identify and remove many of the causal ingredients, particularly those that are currently present within the system. In addition, we can also mitigate the consequences of error by improving the defences, barriers and safeguards, particularly those involved in the detection and containment of unsafe acts.

The prerequisite for a resilient organization is a comprehensive safety information system. This has two components: reactive outcome measures and proactive process measures. These measures can be combined to identify error-provoking factors in the workplace and the organization. They can also reveal gaps in the defences.

MEDA (Maintenance Error Decision Aid) was presented as an example of an effective reactive outcome tool that enables analysts to trace the systemic origins of adverse events. MESH (Managing Engineering Safety Health) was discussed as a maintenance-focused representative of a class of proactive process measures designed to identify and guide the remediation of systemic weaknesses before they are implicated in adverse events. The chapter concluded with a set of questions designed to identify error-detection and error-containment weaknesses in the defences.

The next chapter deals with the over-arching issue of organizational culture. Culture, and particularly safety culture, is something that can have a consistent effect on all aspects of a maintenance system, for good or ill. Getting the culture right is perhaps the single most important part of safety management.

Notes

1 This is not a management textbook and our intention is not to provide general advice on how to manage maintenance operations. There is no shortage of

guidance available on safety management; see, for example, R.H. Wood, *Aviation Safety Programs: A Management Handbook* (2nd edn) (Englewood, CO: Jeppesen, 1997). The relevant international standards should also be familiar to maintenance managers. See also ISO 9001:2000: *Quality Management Systems—Requirements* (Geneva: International Organization for Standardization, 2000) and *AS/ NZS 4360 Risk Management* (Sydney: Standards Australia, 1999). More will be said about quality management in Chapter 12.

2 Boeing, *Maintenance Error Decision Aid* (Seattle, WA: Boeing Commercial Airplane Group, 1994).

3 See J. Moubray, *Reliability-centered Maintenance* (New York: Industrial Press, 1997).

4 J. Reason, *Comprehensive Error Management in Aircraft Engineering: A Manager's Guide* (London Heathrow: British Airways Engineering, 1995). See also Chapter 7 (pp. 125–55) in J. Reason, *Managing the Risks of Organizational Accidents* (Aldershot: Ashgate, 1997) for further details of MESH and its predecessors.

11 Safety Culture

What is a Safety Culture?

Safety culture is a term that nearly everyone uses, but few can agree upon its precise meaning, or how it can be measured. The social science literature offers an abundance of definitions, which is not particularly helpful, but taken together they suggest that the elements of a safety culture can be sub-divided into two parts.[1] The first comprises the beliefs, attitudes and values—often unspoken—of an organization's membership regarding the pursuit of safety. The second is more concrete and embraces the structures, practices, controls and policies that an organization possesses and employs to achieve greater safety.

Rather than striving vainly for a single comprehensive definition, we prefer to emphasize some of the more important attributes of a safe culture. They are listed below.

- A safe culture is the 'engine' that continues to drive the organization towards the goal of maximum attainable safety regardless of current commercial pressures or who is occupying the top management posts. The commitment of the Chief Executive Officer and his or her immediate colleagues exerts a powerful influence upon a company's safety values and practices, but top managers come and go, and a truly safe culture must endure despite these changes.
- A safe culture reminds the organization's members to respect the operational hazards, and to expect that people and equipment will fail. It accepts these breakdowns as the norm and develops defences and contingency plans to cope with them. A safe culture is a wary culture, one that has a 'collective mindfulness' of the things that can go wrong.
- A safe culture is an informed culture, one that knows where the 'edge' is without having to fall over it. This no easy task in industries with relatively few bad events.

- An informed culture can only be achieved by creating an atmosphere of trust in which people are willing to confess their errors and near misses. Only in this way can the system identify its error-provoking situations. Only by collecting, analysing and disseminating information about past events and close calls can it locate where the boundaries between safe and unsafe operations might lie. Without such a corporate 'memory' the system cannot learn.
- An informed culture is a just culture that has agreed and understood the distinction between blame-free and culpable acts. Some unsafe acts will warrant disciplinary action. They are likely to be very few, but they cannot be ignored. Without a just culture, it is difficult, if not impossible, to establish an effective reporting culture.
- A safe culture is a learning culture in which both reactive and proactive measures (see Chapter 10) are used to guide continuous and wide-reaching system improvements rather than mere local fixes. A learning culture is one that uses the discrepancies that inevitably arise between what it intends and what actually happens to challenge its basic assumptions—and has the will to change them when they are shown to be maladaptive.

This list of attributes makes it clear that a safe culture has many interlocking parts. The main components—or, at least, the ones that we will consider in more detail below—are shown in Figure 11.1.

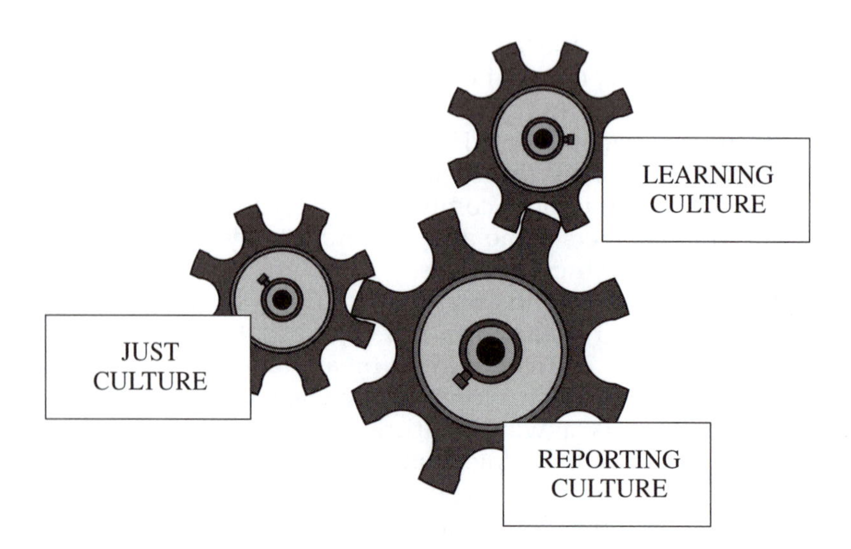

Figure 11.1 The main sub-components of a safe culture

Can a Safer Culture be Engineered?

The answer is yes, up to a point, but it all depends on which features of culture you set out to change. We stated earlier that scientific definitions present culture as having two aspects: what an organization *is* (beliefs, attitudes and values), and what an organization *does* (structures, practices, policies and controls). Faced with a choice, most organizations would seek to change the former through various motivational measures: fear appeals, sticks and carrots (mostly sanctions), making an example of those who commit unsafe acts, naming, blaming, shaming, retraining—and, of course, that old managerial knee jerk reaction, writing another procedure. This is in keeping with the widespread *person model* of human error: the idea that errors and violations arise solely out of the perversity and unreliability of human nature.[2] Since the larger part of this book has been devoted to arguing against such a view, it will come as no surprise that we favour the latter course—changing practices. And here are some of the reasons why.

Geert Hofstede, an expert on culture, expressed the case well when he wrote that: 'Changing collective values of adult people in an intended direction is extremely difficult, if not impossible.'[3] Values, beliefs and attitudes can be changed, but not necessarily by a direct assault. It is often better to approach the issue obliquely by setting out to change an organization's practices. Although there is a widespread view that mental states like attitudes and beliefs drive behaviour, it can also work the other way round. Introducing practices and structures that are seen to work effectively have a way of bringing people's values into line with them.

Consider smoking, for example. In the 1960s, when cigarette smoking was commonplace, there was medical evidence to indicate a strong correlation between tobacco usage and a higher likelihood of lung cancer and heart disease. But almost the only people who changed their behaviour at that time were the doctors who saw the lung tumours on the X-rays, or who had to treat cardiovascular diseases. Now—at least in many countries—smoking is very much a minority habit. What has changed? Did those people who quit smoking believe that they were likely to shorten their lives unless they gave up? Perhaps some of them did. But for the most part, smokers found themselves increasingly marginalized by societal practices that restricted the places where the habit could be indulged. It is still possible to light up, but only in the most awful places—like standing in the rain outside the office doorway or by huddling into the ghastly smoke-filled boxes set aside for the unredeemed in airports and other public areas. The effect was to make committed smokers feel like pariahs, an underclass. In the end, the pleasures of smoking were simply not worth the social costs. It was easier to quit than to listen to the righteous lectures of the recently converted, or to put up with the few unpleasant places in which smoking was still permitted. Altered practices and the increasing social pressure were the main instruments of change.

Creating a Just Culture

Creating a just culture—it could just as well be called a trust culture—is the critical first step in socially engineering a safe culture. No one is going to confess his or her errors and near misses in a wholly punitive culture. Trust is the essential prerequisite for a reporting culture—and hence an informed culture. Nor is a wholly blame-free culture feasible since some unsafe acts, albeit a very small proportion, are indeed culpable and deserve severe sanctions. It would be silly to pretend otherwise since the reckless actions of a small minority not only endanger the safety of the system as a whole, but also pose an immediate threat to other maintainers. If these few 'cowboys' are seen to go unpunished, then management loses credibility. But they would also lose it if they failed to discriminate between this minority of egregious acts and the largely blameless errors that comprise more than 90 per cent of the total number committed. A just culture hinges critically on a collectively agreed and clearly understood distinction being drawn between acceptable and unacceptable behaviour. But how should this distinction be made?[4]

A natural instinct is to draw the line between errors and violations. Errors are largely unintended, whereas most violations have an intentional component. Criminal law makes the distinction between a 'guilty act' (*actus reus*) and a 'guilty mind' (*mens rea*). Except in situations of absolute liability (for example, shooting a red light), neither by itself is sufficient to secure a conviction: to achieve a guilty verdict both the act and the intention to commit the act must usually be proved beyond reasonable doubt. At first sight, therefore, the decision appears to be fairly straightforward, simply determine whether the unsafe act involved non-compliance with safe operating procedures. If it did, then it would be culpable. Unfortunately, the matter is not quite so simple, as the following three scenarios will illustrate.

In all three cases, an aircraft maintainer is asked to inspect the fuselage of an aircraft for any cracked rivets that could jeopardize its airworthiness. The company procedures require that maintainers perform this inspection using an appropriate work stand and an array of lights.

- *Scenario A.* The maintainer collects the work stand and lights from the stores and carries out an approved inspection. But he erroneously fails to spot a cracked rivet.
- *Scenario B.* The maintainer decides not to bother with the work stand and approved lights. Instead, he conducts a cursory examination by walking underneath the aircraft with a flashlight. He too misses a cracked rivet.
- *Scenario C.* The maintainer goes to the store to fetch the work stand and the lights, but finds the former missing and the latter unserviceable. Conscious that the aircraft will soon be in service, he takes a flash light and carries out his inspection from beneath the aircraft. He too misses the cracked rivet.

Note that the error is the same in all three cases: a cracked rivet is not spotted. It is clear, however, that the underlying behaviour was quite different in these three cases. In Scenario A, the engineer was compliant with the procedures. In Scenario B, the engineer could not be bothered to comply. In Scenario C, the engineer intended to comply, but equipment deficiencies made it impossible. Violations were committed in both B and C, but the underlying motivation was clearly not the same. In B, the engineer deliberately took a short cut that increased the probability of missing a cracked rivet. In C, the engineer sought to do the best he or she could with the inadequate materials at hand.

The moral of these three scenarios is clear. Neither the error nor the mere act of violating is sufficient to warrant being labelled as unacceptable. To establish culpability, it is necessary to review the sequence of behaviour of which the error or violation was part. The key issue is did the individual deliberately and without good cause engage in a course of actions likely to promote error?

We have tried to show that the simple distinction between errors and violations can be misleading. Non-compliance may provide a clue to unacceptable behaviour, but it is not sufficient to establish it. Many violations, as Scenario C demonstrated, are system-induced, the product of inadequate tools and equipment, rather than a desire for an easier way of working on the part of the individual. It is also the case that manuals and procedures can be incomprehensible, unworkable, unavailable or just plain wrong. Where these deficiencies are apparent, sanctioning the non-compliant individual will not enhance system safety. At best, it will simply be seen as an indication of management's shortsightedness and induce a sense of 'learned helplessness' in the workforce; at worst, it will be viewed as a symptom of management malevolence. In neither case is an atmosphere of trust and respect likely to flourish.

Can the law help? For negligence, largely an issue for civil law, the enquiry begins with some human action that has brought about a bad consequence. The question that then arises is 'Was this an outcome that a "reasonable and prudent person" could have foreseen and avoided?' Recklessness, a matter for criminal law, involves taking a deliberate and unjustifiable risk. The distinguishing feature is the intention. The actions do not need to have been intended in the case of negligence, but it is necessary to prove prior deliberation to achieve a conviction for recklessness.

How can one make sensible decisions about culpability without a law degree? There are two rule-of-thumb tests that can be applied to each serious event in which unsafe acts were significant contributors.

The Foresight Test

Did the individual knowingly engage in behaviour that an average maintainer would recognize as being likely to increase the probability of making a safety-critical error? In any of the following situations, the answer to this question might well be 'yes', and hence indicative of culpability.

- Performing maintenance under the influence of a drug or substance known to impair performance.
- Clowning around whilst driving a towing vehicle or forklift truck or whilst handling other potentially damaging equipment.
- Becoming excessively fatigued as a consequence of working a double shift.
- Taking unwarranted short cuts like signing off on jobs before they are completed.
- Using tools, equipment or parts known to be sub-standard or inappropriate.

In any of these situations, however, there could be extenuating circumstances. In order to resolve these issues, it is necessary to apply the substitution test.

The Substitution Test

This involves performing a mental test in which we substitute the individual actually concerned in the event for someone else doing the same kind of work and possessing comparable training and experience.[5] The question then is 'In the light of the prevailing circumstances, would this other person have behaved any differently?' If the answer is 'probably not' then apportioning blame has no role to play and would probably obscure the underlying systemic deficiencies. A further question can be put to a group of the erring person's peers: 'Given the situation in which the event occurred, could you be sure that you would not have committed the same or a similar type of unsafe act?' If the answer is again 'no' then blame is very likely to be inappropriate.

Getting the Balance Right

The policy being advocated here has two elements: on the one hand, there is 'zero tolerance' for reckless conduct, but this is linked, on the other hand, to creating a widespread confidence that the vast majority of unintended unsafe acts will go unpunished. Sacking or prosecuting truly reckless offenders, particularly where there is a prior history of similar offences, not only makes the working environment safer, it also sends out a clear message regarding the consequences of unacceptable behaviour. This encourages the vast majority of the workforce to perceive the organizational culture as being just in that it knows the difference between bad acts and honest errors. Natural justice works two ways. Punishments for the few can protect the innocence of the many. It will also encourage the latter to come forward with accounts of errors and near misses. As we shall see later, one of the defining characteristics of high reliability organizations is that they not only welcome such reports, but also commend, even reward, the reporters.

Creating a Reporting Culture

Even with a just culture, there are still a number of psychological and organizational barriers to be overcome before a reporting culture can be created. The first and most obvious is a natural disinclination to confess one's blunders—we do not want to be held up to ridicule. The second is the suspicion that such reports might go on the record and count against us in the future. The third is scepticism. If we go to the trouble of writing an event report that reveals system weaknesses, how can we be sure that management will act to improve matters? Fourth, actually writing the report takes time and effort. Why should we bother?

Below are some of the characteristics of successful reporting programmes.[6] Each feature is designed to overcome one or other of the barriers outlined above.

- *De-identification*. How this is achieved depends on the culture of the organization. Some people prefer complete anonymity—though this has the disadvantage of making it difficult to seek further information to fill in the gaps in the story. Other organizations are content with confidentiality, where the reporter is known only to a very few people.

- *Protection*. Organizations with successful schemes generally have a very senior manager issue a statement guaranteeing that any reporter will receive at least partial indemnity against disciplinary procedures.[7] This usually requires the report to be made within some specified time after the event. It is not possible to offer complete immunity to sanctions since—as we have discussed at length above—some acts are indeed culpable. However, the experience of these successful programmes suggests that such circumscribed guarantees are sufficient to elicit large numbers of reports of honest errors.
- *Separation of functions*. Successful programmes separate the agency or department collecting and analysing the reports from those bodies with the authority to initiate disciplinary proceedings.
- *Feedback*. Successful programmes recognize that reports would soon dry up if the workforce felt they were sending them into an organizational black hole. Rapid, useful, accessible and intelligible feedback to the reporting community is essential. This is often achieved by publishing summary reports of the issues raised and the counter-measures that have been implemented. Some large organizations, like NASA, also maintain databases that are available to legitimate researchers and analysts.
- *Ease of making the report*. Some organizations started their programmes with forms that asked a limited number of forced-choice questions in which respondents were required to indicate which of the listed error types or environmental conditions were relevant to their events. Experience soon showed, however, that reporters prefer a more open and less constrained format. Later versions of the report form encouraged free text reporting in which respondents were able to tell a fuller story and express their own perceptions and judgements. Stories are an excellent way of capturing the complex interaction of many different factors. Such accounts may take longer to complete, but reporters prefer them, particularly since they provide opportunities to express their own ideas about what should be done to prevent a recurrence.

There is no one best way to create a successful reporting system. The list above merely indicates some of the features that have overcome the barriers to filing useful event reports. Each organization must be prepared to experiment to discover what works best.

Perhaps paradoxically, a successful scheme is often seen as one that attracts a steadily growing volume of event reports. While this is likely to be a valid interpretation in the early stages of a programme, logic dictates that a time will come when an increase in the number of event reports cannot simply be taken as a sign of greater trust—it could also mean that a larger number of safety-critical events are occurring within the system. To the best of our knowledge, however, very few, if any, reporting programmes have yet reached this level of discrimination. In any case, the mere volume of reports is never going to be a good index of system safety. It is likely that the numbers alone will always underestimate the true incidence of errors and near misses. The greatest value of a safety information system lies in its ability to identify recurrent event patterns, error traps and gaps or weaknesses in the defences.

Creating a Learning Culture

The most important prerequisite for a learning culture is a reporting culture. Without an effective incident and near miss reporting scheme that collects, analyses and disseminates safety-related information, particularly that concerning the location of safety or quality 'black spots', the organization is not only unaware of the risks, but also lacks a memory. But even when these elements are present, it is still necessary for the organization to adopt the most appropriate learning mode. The social science literature has identified two distinct types of organizational learning: single-loop and double-loop learning.[8] They are illustrated in Figure 11.2.

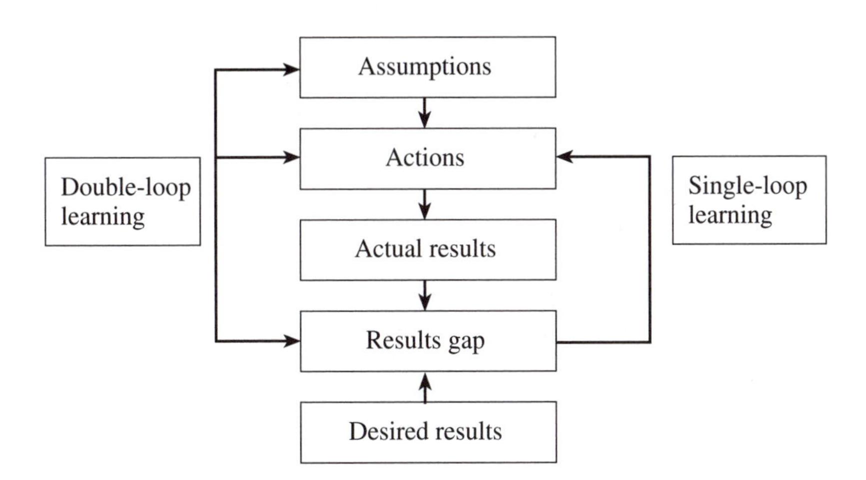

Figure 11.2 Distinguishing between single-loop and double-loop organizational learning

Source: After J.S. Carroll, J.W. Rudolph and S. Hatakenaka, 'Organizational learning from experience in high-hazard industries: Problem investigations as off-line reflective practice', *Research in Organizational Behaviour*, in press, 2001.

The central boxes in Figure 11.2 show the sequential stages involved in organizational action. They begin with the organization's basic assumptions about how things work and how things are done. This 'mental model' shapes the goals and the actions. Once the actions are performed, it is necessary to see whether the actual results conform to the desired results. When there is a discrepancy—the results gap—some modifications need to be made to either the actions or their underlying assumptions. In single-loop learning, only the actions are reviewed. This is the course that organizations subscribing to the 'person model' of human error are most likely to follow. They look for deviant performance on the part of the people who carried out the actions, and its discovery is viewed as the 'cause' of the discrepant results. The learning process stops once person-focused counter-measures (naming, blaming, shaming, retraining and writing another procedure) have been applied. This is single-loop learning and, like the person model of error, it is very widespread. The results of the 'learning' are likely to be restricted to engineering 'retro-fixes' and disciplinary action.

Double-loop learning goes one very important stage further. It not only reviews the prior actions but also challenges the organizational assumptions that prompted them. Double-loop learning leads to global reforms rather than local repairs, and to the adoption of a 'system model' of human error that is concerned not so much with who blundered, but how and why the organization's policies, practices, structures, tools, controls and safeguards failed to achieve the desired result.

John Carroll of the MIT Sloan School of Management has made an extensive study of how a number of high-hazard organizations learn from experience—in particular, from the work of problem or event investigation teams. He and his colleagues produced the four-stage model of organizational learning that is summarized below.[9]

- *Local stage.* Learning is mainly single-loop and is based primarily on the skills and experience of specific work groups. Behaviours may be adjusted after comparison with performance standards, but the underlying assumptions are not challenged. Learning is constrained by a tendency to deny any wider systemic problems and by limited expertise.
- *Control stage.* Many modern industries seek to improve operations and avoid bad events through bureaucratic controls such as sanctions, incentives, standard operating procedures and formal routines. Strenuous efforts are made to limit variation and to avoid surprises. There is a strong preference for single-loop local fixes. Many organizations believe in strict compliance to the rules and have a relatively simple view of cause and

effect. Such an operational style is often successful in stable environments, but it is ill suited to an uncertain, dynamic and turbulent world.

- *Open stage*. The spur to moving into this early stage of double-loop learning is often the need to accommodate widely differing views about the nature of a problem and its solutions. At first, these divergent views may be seen as disruptive, and efforts are made to bring the outliers into line. But eventually it is recognized that the world is not a simple place and that each of these differing views has some validity. Managers often find this a very uncomfortable stage, but recognition of its importance can allow the challenging of cherished assumptions and the development of new and more adaptive ways of working.
- *Deep learning stage*. This is marked by an increasing tolerance for short-term difficulties (that is, the discomfort produced by discrepant views) and greater resources are allocated to the learning process. There is awareness that problems are not someone's fault, rather they are an inevitable feature of any complex system. Managers do not see themselves as 'controlling' the workforce. They see their job as providing the resources necessary for adequate performance. Assumptions are constantly under review. There is a deep-rooted respect for the way that operational hazards can damage the system. This leads to intelligent wariness rather than helpless paralysis. Bad things are expected and planned for. In summary, a deep learning organization has both the will and the resources to strive for continuous improvement.

Types of Safety Culture: The Good, the Bad and the Average

Ron Westrum, an American social scientist, has identified three kinds of safety culture: *generative*, *pathological* and *bureaucratic* (or *calculative*).[10] They echo many of the learning features discussed above. A major distinguishing feature is the way that an organization handles safety-related information.

- *Generative organizations* are characterized by deep learning. They 'encourage individuals and groups to observe, to inquire, to make their conclusions known; and where observations concern important aspects of the system, actively to bring them to the attention of higher management'. We will discuss these attributes in more detail in the next chapter.
- *Pathological organizations* muzzle, malign or marginalize whistle-blowers, shirk collective responsibility, punish or cover up

failures, discourage new ideas and keep one step ahead of the regulator.

- *Bureaucratic organizations*—the large majority—lie somewhere in between. They will not necessarily shoot the messenger, but new ideas often present problems. They tend to be stuck in the control stage of learning and rely heavily on procedures to minimize performance variation. Safety management tends to be isolated rather than generalized, and treated by local fixes rather than by systemic reforms.

Patrick Hudson of Leiden University[11] has extended this three-part classification into five stages, each of which needs to be passed through before the next level can be achieved. Progress through these levels is achieved by growing trust, an increasing informedness and a willingness to engage in double-loop learning.

- Pathological ('who cares as long as we don't get caught').
- Reactive ('safety is important; we do a lot every time we have an accident').
- Calculative ('we have systems in place to manage all hazards').
- Proactive ('we work hard on the problems we still find').
- Generative ('we know that achieving safety is difficult; we keep brainstorming new ways in which the system can fail and have contingencies in place to deal with them').

The most difficult step forward is between the proactive and generative stages. Many proactive organizations are tempted to rest on their laurels, but truly generative organizations know that there are always novel scenarios of system failure. They also understand that a period of time without an event is not good news, merely no news.

Summary

This chapter began with a list of the principal characteristics of a safe culture. A safe culture was shown to have a number of interlocking parts: a just culture, a reporting culture and a learning culture.

Then we posed the question: to what extent can these components of a safe culture be socially engineered? We argued that attempting to change an organization's practices was more likely to be effective than a direct frontal assault on attitudes, beliefs and values. Effective practices—particularly in a technical organization—will eventually bring attitudes and beliefs into line with them. Acting and doing—and getting results—lead to thinking and believing, rather than the other way around.

Trust is the first necessity for a safe culture. But trust and respect cannot exist in either a punitive or a so-called 'blame-free' culture. While the vast majority of unsafe acts involve 'honest' or non-culpable errors, we cannot escape the fact that a small minority commit reckless unsafe acts—and will continue to do so if left unchecked. In order to create a just culture, it is necessary to make a collective agreement as to where the line should be drawn between acceptable and unacceptable errors. We pointed out that the distinction between unintended errors and deliberate violations was insufficient for this purpose, as some violations are committed out of necessity rather than bad intentions. In order to assist in differentiating between blameworthy and blameless acts, we proposed two common-sense tests: the foresight test and the substitution test.

Next we identified some of the psychological and organizational barriers that stand in the way of an effective reporting culture, and then described the steps that successful reporting programmes had taken to overcome them. These features were not intended as a standard prescription, rather they were meant as guidelines to be tailored to the needs of your particular organization.

In discussing the steps involved in creating the groundwork for a learning culture, we distinguished between two kinds of organizational learning—single-loop and double-loop learning. We described a progression of stages that were involved in achieving double-loop learning in which deviations between actual and desired performance lead to a review not only of the immediately preceding actions, but also of the assumptions that guided them.

The final section outlined the characteristics of three types of safety culture—pathological, bureaucratic and generative—and then described the stepwise progression involved in achieving a generative culture. In the next chapter, we look in more detail at the defining characteristics of such a culture. This last chapter also deals with the most difficult part of error management: managing it so that it endures and succeeds.

Notes

1 See J. Reason, 'Achieving a safe culture: theory and practice', *Work and Stress*, **12**, 1998, pp. 293–306; N. Thompson, S. Stradling, M. Murphy and P. O'Neill, 'Stress and organisational culture', *British Journal of Social Work*, **26**, 1996, pp. 647–65; R.L. Helmreich and A.C. Merritt, *Culture at Work in Aviation and Medicine* (Aldershot: Ashgate, 1998).
2 J. Reason, *Managing the Risks of Organizational Accidents* (Aldershot: Ashgate, 1997).
3 G. Hofstede, *Cultures and Organizations: Intercultural Cooperation and its Importance for Survival* (London: Harper Collins, 1994), p. 199.

4 Reason, 1997, op. cit. David Marx, an engineer and lawyer specializing in creating just cultures, was the inspiration for much of the discussion in this section.
5 N. Johnston, 'Do blame and punishment have a role in organizational risk management?', Flight Deck, Spring 1995, pp. 33–6.
6 M. O'Leary and S.L. Chappell, 'Confidential incident reporting systems create vital awareness of safety problems', *ICAO Journal*, **51**, 1996, pp. 11–13; S.L. Chappell, 'Aviation Safety Reporting System: program review', in *Report of the Seventh ICAO Flight Safety and Human Factors Regional Seminar,* Addis Ababa, Ethiopia, 18–21 October 1994, pp. 312–53; M. O'Leary and N. Pidgeon, 'Too bad we have to have confidential reporting programmes', *Flight Deck*, Summer 1995, pp. 11–16.
7 J.A. Passmore, 'Air safety report form', *Flight Deck*, Spring 1995, pp. 3–4.
8 C. Agyris and D. Schön, *Organizational Learning II: Theory, Method and Practice* (Reading, MA: Addison-Wesley, 1996).
9 J.S. Carroll, J.W. Rudolph and S. Hatakenaka, 'Organizational learning from experience in high-hazard industries: Problem investigations as off-line reflective practice', *Research in Organizational Behavior*, in press, 2001.
10 R. Westrum, 'Cultures with requisite imagination', in J.A. Wise, V.D. Hopkin and P. Stager (eds), *Verification and Validation of Complex Systems: Human Factors Issues* (Berlin: Springer-Verlag, 1992), pp. 401–16.
11 P. Hudson, *Aviation Safety Culture* (Leiden: Centre for Safety Science, Leiden University, 2002).

12 Making it Happen: The Management of Error Management

The purpose of this chapter is to pull together the various strands of this book by focusing upon the management of error management. Unless you manage error effectively, error will manage you[1]—and this is especially true of maintenance organizations in which the activities (as indicated at the outset) possess an unusually large number of error-provoking features. The costs of failure are enormous. In Chapter 1, we listed several major accidents in which maintenance failures were a major contributor. Regardless of these catastrophic possibilities, however, many maintenance organizations haemorrhage vast sums of money annually as the result of poorly managed errors. Even if disasters seem a remote possibility, easing the severe pain in the bottom line must surely be a powerful incentive to get the management of error right.

Here Comes Another One

One possible barrier to effective error management is the fact that the managers of maintenance facilities have been required to implement a number of management 'systems' over the past 10 years or so. Is error management just another of these systems, yet another burden to be added to the already heavy workload? In order to convince you that this is not the case, we need to start by sorting out the differences and overlaps between quality management systems, safety management systems and error management.

Quality Management Systems

The Total Quality Management (TQM) movement had its origins in Statistical Process Control (SPC), a technique invented at the Bell

Laboratories in New York in the 1920s.[2] SPC required quality measurements to be made at the point of manufacture rather than at the end of the production line. The quality guru W. Edwards Deming took SPC to Japan in the 1950s—with results that are now widely known—and later re-imported it to the United States in the 1970s where other quality experts such as Juran, Feigenbaum and Ishikawa were having a profound impact on American management. TQM has subsequently spread throughout the industrialized world and has been extensively adopted by aviation maintenance organizations in particular. The key feature of TQM is that quality is everybody's responsibility. It is not something to be 'controlled' at the end of the line by specialist inspectors, but something that has to be 'assured' throughout the entire work process. Hence, we talk of *quality assurance* rather than *quality control*. Quality, in short, must be 'engineered' into the product at every stage.

Quality assurance (QA) is the audit arm of TQM. Its main features are summarized below.

- QA is about assuring customers and others that a system can deliver products and services to the required quality.
- QA achieves this by documenting the way things are supposed to be done, and then auditing to ensure that they are done as intended.
- Any discrepancies are 'fed back' so that the organization can take corrective action, thus continuously improving its performance.

Within the maintenance world, the philosophy of Reliability Centred Maintenance (RCM) has some similarities to TQM. RCM is a principled approach to determining maintenance requirements and has been applied extensively in most sectors of industry.[3]

We will return to Quality Assurance shortly. In the meantime we need to outline the main features of safety management systems.

Safety Management Systems

Safety management systems—within the UK, at least—have their origins in the Health and Safety at Work Act (HSW Act) 1974. It was this Act, framed by the Robens Committee, which initiated the now widespread move towards self-regulation. Prior to that, safety legislation had become extremely fragmented and almost entirely rule-based, imposing heavy obligations upon employers while workers' participation was ensured mainly through disciplinary measures. Unlike previous legislation, the HSW Act did not go into any great detail with regard to specific dangers (for example, machinery, hoists, ladders, lifts and the like); rather it provided broad guidelines for the duties of employers, employees, suppliers and users. In short, safety like quality became everybody's responsibility.

The shift from rule-based to more goal-based regulation was subsequently given a huge push by the Cullen Report on the *Piper Alpha* disaster in 1988.[4] This was published in 1990 and created a new regulatory model for offshore oil and gas installations—that later spread across a wide range of hazardous industries. The new model required operators to undertake a formal safety assessment of their major hazards and to demonstrate that suitable controls, defences and safeguards had been put in place. The product of this exercise was termed the Safety Case. A Safety Management System (SMS) was the major component of the Safety Case. To quote from the Cullen Report:

> The Safety Management System (SMS) should set out the safety objectives, the system by which these are to be achieved, and the performance standards which are to be met and the means by which adherence to these standards is to be monitored. It should draw upon quality assurance principles similar to those in ISO 9000.[5]

Among the oil and gas companies, Shell Exploration and Production took the lead in the development of the first generation of safety management systems. These pioneering efforts have left their stamp on most subsequent systems. The safety management process contains four basic steps:[6]

- *Hazard identification*. What are the hazards encountered in our operations? What can go wrong?
- *Risk assessment*. How dangerous is the hazard? What is the likelihood that it will cause harm or losses?
- *Defences and safeguards*. What defences are needed to warn, protect against and contain the harmful consequences of these hazards?
- *Recovery*. What must be done if something goes wrong?

In keeping with the ISO 9000 recommendation, each of these steps must be documented—this is the 'system' component of SMS, and should include the following elements.[7]

- *Safety management policies*. These identify the top-down processes by which top management express their commitment to achieving the organization's safety goals and define the organization's safety policies.
- *Safety management principles*. These are derived from the policies, and should specify the safety objectives by which the

organization intends to demonstrate its compliance with its policy statements.

- *Safety management processes.* The organization should write procedures describing the means by which the safety objectives are to be met. These should include statements relating to controls, accountabilities, safety-critical activities and competencies.
- *Safety assurance documentation.* These should cover the current operations, and embrace all the units, facilities and activities within them. The purpose is to provide assurance that these entities are safe for continued operation. They should also include documentation covering any proposed changes or additions to existing systems. Safety assessments should be carried out prior to the introduction of any new system or a change to an existing system.

Once completed, the Safety Case—comprising a formal safety assessment, the safety-ensuring programmes and the documentation of the policies, principles and processes—becomes the standard against which the regulator determines safety compliance for that particular organization. This move to goal-based regulation has brought a number of advantages, particularly the necessity for the regulated organizations to think for themselves, often for the first time, about the hazards that beset their operations. Producing a satisfactory SMS also requires that the organization understands exactly how it does its business—sometimes for the first time. But there are problems, as we shall discuss shortly.

The Common Features of Safety and Quality Management Systems

Given that the quality management movement markedly influenced the introduction of safety management systems, it is hardly surprising that they share a number of common beliefs and features.

- Neither quality nor safety can be achieved by a piecemeal approach. Both need to be planned and managed.
- Both rely heavily on measurement, monitoring and documentation.
- Both involve the whole organization: every function, every process and every person.
- Both strive for small continuous improvements rather than dramatic advances.

But they have shared problems as well. Because of their extensive reliance on documentation—both are sometimes (though not always) massively paper-intensive systems—there is the risk that form can overwhelm substance and that a multitude of fat ring-binders (to say nothing of the expensive man hours that went into producing them) can be regarded as demonstrating the actual presence of quality or safety within the organization. But these things reside more in the views of top managers, the organizational culture, the mindsets and work practices of the frontline operators than in these paper tokens, no matter how voluminous they might be. Regulators, managers and auditors are very busy people, and there is always the temptation to use a checklist approach when evaluating either type of system. If the appropriate sections of the documentation are present and correct, there is a natural tendency to assume that what they are supposed to signify actually exists in the real world.

If this seems unduly harsh, consider the following 'quality-assured' maintenance accidents: the BAC 1-11 windscreen blow out in 1990;[8] the event in which spoilers were left in the maintenance mode in an A320 in 1993;[9] and the loss of engine oil due to missing engine rotor drive covers in a Boeing 737-400 in 1995.[10] In all three cases, the maintenance organizations had moved from quality control to quality assurance; in each event a senior aircraft engineer signed off his own work as being satisfactory when it was not; and all occurred during the night shift when no member of the QA department was available to assess the adequacy of the working practices. Another quality-assured accident was the 1994 Moura mine disaster in Queensland in which 11 men died in a methane gas explosion.[11] Earlier in the same year, the Moura mine was accredited as being quality assured under AS3902.

There are two important lessons to be learned from these events. First, there must be appropriate systems in place to audit—that is, ones that properly identify and control the hazards, particularly the human risks. Second, it is no good relying on the 'paper trail' if it does not correspond to what actually happens. The smart auditor should always be imagining the various ways in which the system could fail. The key question for him or her is 'Would all of these possible system failures have been revealed by this audit method?'

There is one last difficulty with quality and safety management systems: neither starts from the now well-established fact that human and organizational problems—rather than technical or engineering failures—now dominate the risks to maintenance organizations. Quality management gurus, being people of their time (the 1950s to the 1970s), regarded error as something due to carelessness, or to some comparable perversity of human nature. Safety management systems are essentially regulatory instruments, and regulators tend to have either a technical or an operational background. Neither of these parties is especially well acquainted with the developments that have taken place in the behavioural and social sciences over the past 20 years or so. And this is why error management has an essential role to play—and why the measures discussed in this book need to become an integral part of maintenance management.

Why Error Management is Necessary

Since error poses a serious threat to quality and safety, its control should play a major role in both of these management systems. Indeed, if either of them did not address the control of error, it would be seriously deficient. But in our experience error management does not feature as prominently as it should in either of these systems. The dominant emphasis in most existing quality and safety management systems is upon the documentation of largely technical or administrative processes, and this is hardly surprising given the predominantly engineering, operational or managerial backgrounds of their creators. The main aim of this book has been to provide the background knowledge, the models and the tools necessary to begin the process of redressing this imbalance. In order to make this purpose clearer, it is helpful to list the characteristics of error management that distinguish it from the quality and safety management systems discussed above.

- Effective error management derives more from having an appropriate mindset than from extensive documentation. In this sense, it is not a 'system' as such; something that can be 'demonstrated' by the existence of a heavy stack of ring-binders—though it can and should be documented when its elements are incorporated as part of a formal quality or safety management system.
- Effective error management takes Murphy's Law as its starting point: what can go wrong will go wrong. Errors and quality lapses are to be expected: they are a fact of human life like breathing and dying. Though they cannot be eliminated, a wide range of measures aimed at the person, the team, the task, the workplace and the system at large can control errors. There is no one best way. Each organization should apply the measures that best suit its manner of working.
- Effective error management requires an understanding of the varieties of human error and the conditions likely to promote them. Different types of error require different counter-measures. Where possible, however, it is always better to remedy

situations rather than seek to change the human condition. Events and mishaps are more likely to be the result of error-provoking tasks and workplace conditions than of error-prone people. Engineering solutions, either technical or social, often work better than psychological ones, and their effects are longer lasting.

- Effective error management needs an informed culture, one that has a 'collective mindfulness' of the factors that disrupt human performance and knows where the edge lies between productive and dangerous activities. This, in turn, requires the creation of an organizational culture that is just, willing to report its unsafe acts and able to learn from them.
- Simply putting some error-reducing or error-containing tool in place and assuming that it will work all by itself without further attention can never achieve effective error management. This is not an uncommon belief in many technical managers. They are accustomed to installing pieces of equipment and then switching them on with the reasonable expectation that they will do what they are supposed to do so long as the power supply lasts. They are also used to having a list of jobs that can be ticked off after each one has been actioned. But error management tools are not like this. They cannot be ticked off the list. They need to be watched, fussed over, tweaked, massaged and adjusted on a regular basis. This is why we stated earlier that an appropriate mindset is the essence of effective error management.

There is one further important difference between quality and safety management systems and error management. Whereas the former are predominately top-down in their application—imposed either by senior management or by the regulator—error management has a large bottom-up component. The documentation of the quality and safety systems states how the organization ought to function. In sharp contrast, error management tools such as event reporting programmes and proactive process measures (like MESH, for example) reveal how things actually are. One is normative in its approach; the other is descriptive. As such, they are complementary. Error management does not supplant quality or safety management systems. It gives them both an added dimension.

More on Mindset

The American social scientist Karl Weick made the telling observation that reliability is 'a dynamic non-event'.[12] It is dynamic because processes remain under control due to the continuous adjustments, adaptations and compensations made by the human elements of the system. It is a non-event because 'normal' outcomes claim little or no attention. The paradox is rooted in the fact that events claim attention, while non-events, by definition, do not.

Recently, Weick and his co-workers challenged the received wisdom that an organization's reliability depends upon the consistency, repeatability and invariance of its routines and activities. Unvarying performance, they argue, cannot cope with the unexpected. To account for the success of high reliability organizations (HROs) in dealing with unanticipated events, they distinguish two aspects of organizational functioning: cognition and activity. The cognitive element relates to being alert to the possibility of unpleasant surprises and having the collective mindset necessary to detect, understand and recover them before they bring about bad consequences. Traditional 'efficient' organizations strive for stable activity patterns yet possess variable cognitions—these differences are most obvious before and after a bad event. In HROs, on the other hand, flexibility is encouraged in their activity, but there is consistency in the organizational mindset relating to the operational hazards. This cognitive stability depends critically upon an informed culture—or what Weick and his colleagues have called 'collective mindfulness'.[13]

Collective mindfulness allows an organization to cope with the unanticipated in an optimal manner. 'Optimal' does not necessarily mean 'on every occasion', but the evidence suggests that this chronic unease is a critical component of organizational resilience. Since bad events are rare, intelligently wary organizations work hard to extract the most value from what little data they have. They actively set out to create a reporting culture by commending, even rewarding, people for reporting their errors and near misses. They work on the assumption that what might seem to be an isolated failure is likely to come from the confluence of many 'upstream' causal chains. Instead of localizing failures, they generalize them. Instead of applying local repairs, they strive for system reforms. They do not take the past as a guide to the future. Aware that system failures can take a wide variety of yet-to-be-encountered forms, they are continually on the lookout for 'sneak paths' or novel ways in which Murphy (he of the Law) and his partner, Sod, can defeat or bypass the system defences.

In Search of Resilience

By resilience here, we mean those properties of an organization that make it more resistant to its operational hazards. Typically, organizational resilience is assessed by counting the number of adverse events it sustains in a given period of time. But these negative outcome measures are unsuitable for the purpose. In the first place, they may reflect occasional moments of weakness rather than some underlying state of organizational 'health'. Second, chance plays a large part in whether or not bad events occur. But chance works both ways. It can afflict the deserving and protect the unworthy. The absence of bad events during a given time interval does not necessarily mean that an organization is healthy, nor does the occurrence of isolated events signify that it lacks basic robustness. In short, we need another kind of measure.

Engineers use a very direct measure of resilience: the test to destruction. Outside of the workshop, a comparable test is the average number of things that need to go wrong before a system breaks down. A recent study examined 90 fatal accident investigation reports published by the UK Air Accident Investigation Branch between the 1970s and the 1990s with the purpose of establishing how many of 16 possible contributory factors—pilot error, weather, engine failure, airframe problems, insufficient fuel and the like—were cited by the investigators as contributing to the accident.[14] Three types of aircraft were compared: large commercial jets, light aircraft (general aviation) and helicopters. The results were clear. On average, it took 1.95 problems to down a helicopter, 3.38 problems for light aircraft and 4.46 problems for large jets. The latter were clearly more resilient than helicopters or light aircraft.

Although the method described above has obvious face validity, it is difficult to apply to systems that have not suffered a catastrophic event. Another approach is to ask what are the defining properties of a resilient organization? We can obtain a good deal of information about what these are likely to be from looking at the characteristics of the high reliability organizations, discussed earlier.

Presented below are two checklists that seek to identify some of the characteristics of organizational robustness. The first checklist (Human Performance Awareness Checklist—or HPAC) is designed to elicit views from organizational members regarding how sensitive their system is to the origins of human performance problems and the methods that are appropriate for dealing with them.[15] The second checklist (Checklist for Assessing Institutional Resilience—or CAIR) takes a broader approach and is more orientated towards how safety and error management tools are applied in the respondent's organization. Both checklists are built on the assumption that organizational resilience is a product of three C's: commitment, competence and cognizance.

The Three C's

- *Commitment.* In the face of ever-increasing commercial pressures, does top management have the will to make error management and safety management work effectively?
- *Cognizance.* Do your managers understand the nature of the 'safety war'—particularly with regard to human and organizational factors?
- *Competence.* Are your safety and error management tools understood, adequate for the purpose and properly utilized?

Human Performance Awareness Checklist

Respondents are asked to express their beliefs about the extent to which each of 30 statements holds true for their organization. To avoid response bias, approximately half the items are phrased in a positive direction (where agreement is consistent with a resilient organization) and half in a negative direction (where disagreement is consistent with resilience). The scoring direction is shown by a sign (+ or –) in parentheses. Statements where your responses are consistent with organizational resilience score one, a 'don't know' scores zero.

For convenience, the statements set out below are clustered according to the appropriate 'C', although they would be presented to normal users in a scrambled order. The purpose of the list below is to indicate the attributes of resilience rather than to produce a score.

Commitment items

- If something goes wrong, management looks for someone to blame. (–)
- Human performance issues are high on management's agenda. (+)
- Management is only interested in the bottom line. (–)
- When there are human performance problems, managers do their best to fix the conditions that promoted them. (+)
- Managers believe that the procedures are always correct and applicable. (–)
- Managers are genuinely interested in matters relating to human performance. (+)
- Managers fail to recognize unsuitable working conditions that produce recurrent human performance problems. (–)
- Managers often discuss working conditions and human performance problems with people working at the 'sharp end'. (+)

- Management believes that the threat of disciplinary action is the best way to minimize errors. (–)
- Management is willing to act upon good suggestions for improving safety and reliability, even when they come from junior employees. (+)

Cognizance items

- Our human factors (HF) personnel are well trained and keep up with developments in the HF community. (+)
- Managers believe that only frontline maintainers make dangerous errors. (–)
- Managers are more interested in quick fixes than system reforms. (–)
- Our first-line supervisors are trained to a very high level of competence. (+)
- We expect errors to occur and train staff to detect and recover them. (+)
- Managers believe that it is cheaper and easier to change people's behaviour than the working conditions. (–)
- Management does not appreciate that defences and controls can create problems as well as provide protection. (–)
- Our managers and supervisors have a good understanding of the workplace factors that are likely to promote errors and violations. (+)
- Each event in which unsafe acts are implicated is carefully reviewed and the people involved are treated justly. (+)
- Management does not appreciate that our existing procedures cannot cover all eventualities. (–)

Competence items

- If we come up with a safer and/or more reliable way of working, we are given credit for it and the information is widely disseminated. (+)
- We rarely discuss human performance issues before we start a new job or change working conditions. (–)
- We frequently see managers on the shop floor. (+)
- All personnel receive some basic training in human factors issues. (+)
- Employees are reluctant to report errors and near misses because they fear they could be punished. (–)
- When someone is uncertain about how to do a job, there is always someone willing and able to advise them. (+)

- Employees are actively discouraged from raising issues relating to human performance. (–)
- When a serious event occurs, management is more interested in discovering how and why the defences failed than in finding someone to blame. (+)
- We do not have an effective incident and error-reporting programme. (–)
- The same kinds of events keep happening over and over again. (–)

These are not exhaustive lists of items. There are many others that could and should be included—try making up your own lists. However, these items are sufficient to give an idea of the extent of your organization's preparedness for human factors problems and whether or not suitable counter-measures have been put in place. Scores of less than 15 indicate that the organization is vulnerable to losses and disruption due to errors (the maximum organizational resilience score is 30).

Checklist for Assessing Institutional Resilience (CAIR)

CAIR assesses the extent to which the attitudes and practices of an organization match up to a 'wish list' of the features characterizing a resilient system (see Table 12.1). It was conceived by asking how the three C's—commitment, cognizance and competence—might be manifested at each of four levels of managerial application: philosophy, policies, procedures and practices.

The question for the respondent is 'Does your organization have the following properties?' Each item is scored one, a half or zero (see below). A score of 16–20 is probably too good to be true. Scores between 8 and 15 indicate a moderate to good level of intrinsic resistance to your human and organizational hazards. Anything less than five suggests an unacceptably high degree of vulnerability.

Finally, a health warning: good scores on CAIR provide no guarantee of immunity from maintenance mishaps. Even the 'healthiest' organizations can still have bad events. High reliability requires intelligent wariness and a continuing respect for the many ways in which things can go wrong. Complacency is the worst enemy. Human fallibility and latent system weaknesses will not go away, so there will be no final victories; but error and its consequences can be managed.

Table 12.1 Checklist for Assessing Institutional Resilience (CAIR)

Complete the checklist as follows:
Yes = this is definitely the case in my organization (score 1)
? = don't know, maybe or it could be partially true (score 0.5)
No = this is definitely not the case in my organization (score zero)

	Yes	?	No
Managers are ever mindful of the human and organizational factors that can endanger their operations.			
Managers accept occasional setbacks and nasty surprises as inevitable. They anticipate that staff will make errors and train staff to detect and recover them.			
Top managers are genuinely committed to the furtherance of system safety and provide adequate resources to serve this end.			
Safety-related issues and human performance problems are considered at high-level meetings on a regular basis, not just after some bad event.			
Past events are thoroughly reviewed at top-level meetings and the lessons learned are implemented as global reforms rather than local repairs.			
After some bad event, the primary aim of top management is to identify the failed system defences and improve them, rather than seeking to divert responsibility to particular individuals.			
Top management adopts a proactive stance towards safety. That is, it does some or all of the following: takes steps to identify recurrent error traps and remove them, strives to eliminate the workplace and organizational factors likely to provoke errors, 'brainstorms' new scenarios of failure, and conducts regular 'health checks' on the organizational processes known to contribute to accidents.			

Table 12.1 continued

	Yes	?	No
Top management recognizes that error-provoking system factors (e.g., under-manning, inadequate equipment, inexperience, patchy training, bad human–machine interfaces and the like) are easier to manage and correct than fleeting psychological states such as distraction, inattention and forgetfulness.			
It is understood that the effective management of safety, just like any other management process, depends critically on the collection, analysis and dissemination of relevant information.			
Management recognizes the necessity of combining reactive outcome data (i.e., the near miss and incident reporting system) with proactive process information. The latter entails far more than occasional audits. It involves the regular sampling of a variety of institutional parameters (e.g., scheduling, budgeting, rostering, procedures, defences, training and the like), identifying which of these 'vital signs' is most in need of attention, and then carrying out remedial actions.			
Representatives from a wide variety of departments and levels attend safety-related meetings.			
Assignment to a safety-related or human factors function is seen as a fast-track appointment, not a dead end. Such functions are accorded appropriate status and salary.			
Policies relating to near miss and incident reporting systems make clear the organization's stance regarding qualified indemnity against sanctions, confidentiality and the organizational separation of the data-collecting department from those involved in disciplinary proceedings.			

Table 12.1 continued

	Yes	?	No
Disciplinary policies are predicated on an agreed (i.e., negotiated) distinction between acceptable and unacceptable behaviour. It is recognized by all staff that a small proportion of unsafe acts are indeed reckless and warrant sanctions, but that the large majority of such acts should not attract punishment.			
Line management encourage their staff to acquire the mental as well as the technical skills necessary to achieve safe and effective performance. Mental skills include anticipating possible errors and rehearsing the appropriate recoveries.			
The organization has in place rapid, useful and intelligible feedback channels to communicate the lessons learned from both the reactive and the proactive safety information systems. Throughout, the emphasis is upon generalizing these lessons to the system at large rather than merely localizing failures and weaknesses.			
The organization has the will and the resources to acknowledge its errors, to apologize for them and to reassure the victims (or their relatives) that the lessons learned from such accidents will help to prevent their recurrence.			
It is appreciated that commercial goals and safety issues can come into conflict and measures are in place to recognize and resolve such conflicts in an effective and transparent manner.			
Policies are in place to encourage everyone to raise safety-related issues.			
The organization recognizes the critical dependence of effective safety management on the trust of the workforce—particularly in regard to error and incident reporting programmes.			

Summary

This chapter deals with what is perhaps the toughest aspect of error management—making it happen and keeping it going. We began by looking at the differences and similarities between quality and safety management systems and error management. Although error management has a crucial role to play in both quality and safety management systems, it is distinctive in a number of respects. In the first place, it has its origins in recent developments in the study of human and organizational factors. It is based upon an understanding of the varieties of human error and the workplace and system factors that provoke unsafe acts. Error management is not a formal management 'system' in the way that quality and safety management systems are, rather it comprises basic human factors training combined with a range of tools targeted at various levels of the system: the individual maintainer, the maintenance team, maintenance tasks and the maintenance organization as a whole. Error management is bottom-up rather than top-down in its approach. It is based upon the way things are rather than the way they ought to be. As such, it provides an essential complement to both quality and safety management.

Perhaps the most distinctive feature of effective error management, however, is that it depends more on acquiring an appropriate mindset than upon the investment of large amounts of money, employee hours or other costly resources. High reliability organizations, which operate in hazardous conditions yet have fewer than their 'fair share' of accidents, are characterized by 'collective mindfulness'.[16] They maintain a continuing awareness of the possibility of human, technical or organizational failures. They expect that errors will be made and train their personnel to understand, anticipate and recover them. They work hard to create a reporting culture and make the most out of limited event data by generalizing rather than localizing the lessons from past failures. They 'brainstorm' possible event scenarios and have plans in place to cope with them.

The final part of the chapter examined the properties of organizational resilience—a system's ability to withstand the 'slings and arrows' of its operational fortunes. Striving for zero events is an unrealistic goal; the only practicable objective—while still staying in business—is to achieve the maximum degree of resilience and then to sustain it. It was argued that the cultural drivers of enduring resilience were the three C's: commitment, cognizance and competence. This notion was incorporated into two checklists designed to assess an organization's robustness to human factors problems and safety hazards. Their inclusion in this chapter was mainly for the purpose of showing the scope of these features as they might be revealed at various levels of management activity: philosophies, policies, procedures and practices. Neither checklist represents a comprehensive listing of these properties. Readers are encouraged to think of additional indicators of an organization's 'health'.

Some final words of advice: one of the main reasons why error management programmes have faltered in the past is because the champions appointed to implement and oversee them have been posted elsewhere. Try to have your error management programme run from the top—ideally, it should become the CEO's pet project. At the very least, appoint a senior champion who believes in the error management programme and is willing to stick it out when the going gets tough. He or she should report directly to the board. In our experience, the best time to initiate an error management programme is when senior managers have just been seriously frightened by a bad event in which both unsafe acts and organizational factors were implicated. The beneficial effects of such shocks to the system do not last, but they are often sufficient to get the programme rolling. After that it is up to the organizational culture to keep it going.

Do not attempt everything at once. Start with something simple that is likely to have widespread effects. Run a short training programme on human and organizational factors combined with some targeted omission management (see Chapter 9). The Boeing Airplane Company, for example, initiated a successful error management programme with a project dedicated to preventing the omission of lock wires during reassembly. Effective progress in error management is more a matter of evolution than revolution. It does not happen overnight or even in a single year. Think *kaizen*—this is a process of deliberate, patient, continual refinements. Michael Crichton illustrated the meaning of this Japanese word very neatly in his novel *Rising Sun*:

> Americans are always looking for the quantum leap, the big advance forward. Americans try to hit a home run—to knock it out of the park—and then sit back. The Japanese just hit singles all day long, and they never sit back.

Notes

1 We are very grateful to Julian Dinsell for this aphorism.
2 T. Bendell, J. Kelly, T. Merry and F. Sims, *Quality: Measuring and Monitoring* (London: Century Business, 1993).
3 J. Moubray, *Reliability-centered Maintenance* (New York: Industrial Press, 1997).
4 The Hon. Lord Cullen, *Public Inquiry into the Piper Alpha Disaster* (London: HMSO, 1990).
5 Ibid., vol. 2, p. 388.

6 SIPM, *The Safety Management System SIPM Guidance. Vol I*, Report EP 92-0100 (The Hague: Shell Internationale Petroleum Maatschppij BV, 1992).
7 Adapted from Civil Aviation Authority, *Safety Management Systems: SRG Policy and Guidelines* (Gatwick: Civil Aviation Authority, Safety Regulation Group, 1998).
8 See Air Accident Investigation Branch, *Report on the Accident to BAC One-Eleven, G-BJRT, over Didcot Oxfordshire on 10 June 1990*, Department of Transport (London: HMSO, 1992).
9 See Air Accident Investigation Branch, *Report on Incident to Airbus 320-212, G-KMAM at London Gatwick Airport on 26 August 1993*, Department of Transport (London: HMSO, 1995).
10 See Air Accident Investigation Branch, *Report on the Incident to Boeing 737-400, G-OBMM, Near Daventry on 23 February 1995*, Department of Transport (London: HMSO, 1996).
11 A. Hopkins, *Managing Major Hazards: The Lessons of the Moura Mine Disaster* (St Leonards, NSW: Allen and Unwin, 1999). This book also contains an interesting discussion of 'quality-assured' accidents.
12 K.E. Weick, 'Organizational culture as a source of high reliability', *California Management Review*, **29**, 1991, pp. 112–27.
13 K.E. Weick, K.M. Sutcliffe and D. Obtsfeld, 'Organizing for high reliability: Processes of collective mindfulness', *Research in Organizational Behaviour*, **21**, 1999, pp. 81–123.
14 See J. Reason, *Managing the Risks of Organizational Accidents* (Aldershot: Ashgate, 1997).
15 We are exceedingly grateful for the collaboration of John Wreathall in the preparation of this checklist.
16 See also T.R. La Porte and P.M. Consolini, 'Working in practice but not in theory: Theoretical challenges of "high-reliability" organizations', *Journal of Public Administration Research and Theory*, **1**, 1991, pp. 19–47.

Index